A RISK TOO FAR
A PSYCHOLOGICAL AUTOPSY
OF THE PLANNING FOR ARNHEM

A Risk Too Far
A Psychological Autopsy
of the Planning for Arnhem

GARY BUCK

Howgate Publishing Limited

Copyright © 2025 Gary Buck

First published in 2025 by
Howgate Publishing Limited
Station House
50 North Street
Havant
Hampshire
PO9 1QU
Email: info@howgatepublishing.com
Web: www.howgatepublishing.com

All rights reserved.

No part of this publication may be reproduced, stored in a retrieval system, or transmitted in any form or by any means including photocopying, electronic, mechanical, recording or otherwise, without the prior permission of the rights holders, application for which must be made to the publisher.

British Library Cataloguing-in-Publication Data
A catalogue record for this book is available from the British Library

ISBN 978-1-912440-77-1 (pbk)
ISBN 978-1-912440-78-8 (hbk)
ISBN 978-1-912440-79-5 (ebk – ePUB)

Gary Buck has asserted his right under the Copyright, Designs and Patents Act, 1988, to be identified as the author of this work.

The views expressed in this publication are those of the author and do not necessarily reflect official policy or position.

CONTENTS

Figures	vi
Tables	vi
Foreword	vii
Preface	x
Acknowledgements	xix
Abbreviations	xx
Advance to Contact	1
Part One	
1. Montgomery's Observation – Discord	10
2. Montgomery's Orientation – Grip	27
3. Montgomery's Decision – Avoidance	52
4. Montgomery's Action – Dissonance	68
5. Dilemma	82
Part Two	
6. Browning's Observation – Consistency	92
7. Browning's Orientation – Ambition	102
8. Browning's Decision – Bolstering	114
9. Browning's Action – Endowment	128
10. Feasibility	140
Part Three	
11. Urquhart's Observation – Conformity	153
12. Urquhart's Orientation – Complexity	163
13. Urquhart's Decision – Hypervigilance	175
14. Urquhart's Action – Framing	191
15. Risk	210
Withdrawal	216
Bibliography	226
Index	234

FIGURES

P.1	Operation Market Garden	xii
1.1	By September 1944, the Allies had pushed right up to the Siegfried Line	11
12.1	Modes of Complexity	169
14.1	Landing and drop zones, and routes into Arnhem	192

TABLES

5.1	ACH Matrix Template	83
5.2	ACH Matrix – Criteria	86
5.3	ACH Matrix – Criteria and Hypotheses	87
5.4	ACH Matrix – Complete	88
5.5	ACH Matrix – Revised	89
10.1	Scenario Analysis Template	144
10.2	Mainline Scenario	145
10.3	Good Case Scenario	148
10.4	Best Case Scenario	149
15.1	Urquhart's Plan: Outside-In Analysis	212

FOREWORD

Gary Buck and I were born just a few months apart in 1967 entering a world still filled with middle aged men who had seen service in the Second World War. One such man, head teacher Mr Dawson, taught me at primary school and had a strong influence on a boy fascinated by all things military. His office contained a glass fronted cabinet which, although it is astonishing today, proudly displayed a Sten gun and a hand grenade. Whenever Mr Dawson's door was open, I would wander past slowly to stare at the artefacts and occasionally he'd invite me in to regale me about his wartime experiences. I remember nothing of the detail, but I do recall the names of two of his battles: Dunkirk and Arnhem. Operation Market Garden and the Battle of Arnhem were to become of particular interest to me – the boldness of the plan, the race against time, the *en masse* deployment of Allied airborne forces – and I have no doubt the film *A Bridge Too Far* was primarily responsible for my inquisitiveness. Why the remarkable undertaking in September 1944 did not achieve its objective and who was ultimately responsible for the failure, are questions I first sought to answer (very poorly) in my undergraduate dissertation and, despite my many and varied historical interests, intrigue me still.

My opinions about Market Garden have developed significantly over the years. The reasons why can be found in my acquisition of a deeper knowledge of what might be called the 'traditional military history' of events and their causes, but also in gaining a broader academic knowledge from disciplines other than my own. A significant part of that breadth came from an improved understanding about the factors shaping organisational culture and military behaviours while working with experts in the Ministry of Defence. Such was their impact on my thinking that I felt it necessary to completely reassess my approach to my analysis of the past and that demanded I employ new methodologies. Prior to this new realisation, which some colleagues challenged as disloyal absurdity, I was aware of Norman Dixon's influential 1976 book *On the Psychology of Military Incompetence*, but I found it fatally undermined by a very poor grasp of history. It was not, therefore, until I witnessed the outstanding work being done by the British Army's teaching behavioural scientists and embedded occupational psychologists, that I truly embraced a multi-disciplinary approach. From that point my conversion was rapid and having co-found the Centre for Army Leadership (CAL) at the Royal Military Academy Sandhurst in early 2017, immediately endeavoured to develop exceptional leaders across all ranks by drawing on the wisdom proffered by a wide range of outstanding practitioners and academics. Having codified the British Army's leadership philosophy in its first ever 'Leadership Doctrine', the CAL's priority was then to test the document's resilience against experience, knowledge and opinion. Undoubtedly the most instructive event in this process was a battlefield study to Arnhem with participants including every rank from Private to Major General and drawn from a variety of arms. Titled 'Exercise Arnhem Leader' and directed by a team of military historians and psychologists, the four-day visit thoroughly examined the doctrine's key leadership concepts and, from what was learned, improvements were subsequently made to the document. My own intimate understanding of the battle was also challenged by what was a forensic interrogation of events and I found myself checking my views about the behaviour of planners, commanders and leaders and the consequences of their behaviour. It was a humbling experience.

    I had visited Arnhem many times before the CAL exercise, but most memorable was a reconnaissance trip conducted in preparation for a Camberley Staff College tour in 1994 where we were joined by several of the battle's most senior veterans. I was particularly delighted to be teamed up with Geoffrey Powell, a 156 Para company commander in September 1944, author of my favourite wartime memoir *Men at Arnhem*, and a military historian. Together we spent several fascinating days retracing his route across the battlefield, but while in my youthful naivety I was expecting Geoffrey to talk about tactics and fighting methods, he instead focussed on the challenges of effective command, leadership and followership. These were themes readily taken up when joined by the other veterans and members of the military directing staff and soon developed into a discussion about the personalities, influence and effectiveness of key commanders – and a host of passionately held views were shared.

    The vigorous debate the Staff College team enjoyed in the bar of the Hotel Wolfheze in Oosterbeek all those years ago would have been very

Foreword

much better informed had we had access to the research that Gary Buck now presents in this stimulating and informative volume. *A Risk Too Far: A Psychological Autopsy of the Planning of Arnhem* answers many questions I have long had about Bernard Montgomery, Frederick Browning and Roy Urquhart and, once again, I find myself adjusting my views about the Battle of Arnhem. This book is a valuable and much needed contribution to the ever-growing literature on Allied operations in 1944, but more than this, it tells us something about the enduring impact of human psychology on the profession of arms. *A Risk Too Far* is vital reading and, I hope, encourages similar approaches to other battles.

Lloyd Clark
Director of Research, Centre for Army Leadership,
Camberley and Professor of Modern War Studies,
University of Buckingham

PREFACE

Background

This book about Arnhem is different. I want to take an alternative approach and examine the underlying psychological mechanisms that drove the errors made in the planning of Operation Market Garden. Arnhem has fascinated me for a long time. I can still remember seeing the film *A Bridge Too Far* as a ten-year old at the Norwich Odeon when it came out in 1977. It captured my imagination then and I have essentially been its Prisoner of War ever since; it prompted me to read the 1974 book by Cornelius Ryan, from which the film was adapted.

Both the book and the film further fueled my desire to have a full-time career in the British Army, sadly, this was not to be. I have, however, enjoyed a career as a Reservist, with experience serving in Afghanistan, Bosnia and Iraq. Although my professional career turned to psychology rather than the Army, I have remained an avid student of military history and have been lucky enough to combine the two disciplines. This book hopefully represents the fusion of those two interests, and you will, therefore, note two different narrative voices in the discussion that follows as I switch between psychologist and military historian to examine the psychology that underpinned the (flawed) planning for Operation Market Garden, in September 1944.

Overview

Market Garden was the Allied attempt in September 1944 to set the conditions for a thrust into northern Germany and bring the Second World War to an early close by seizing the industrial centre of the Ruhr and then Berlin. The plan involved a strong direct thrust through northern Holland to gain a crossing over the Rhine, the last major physical barrier to the Reich's capital, Berlin. This would have allowed the Allies to turn the flank of the main

German defensive barrier, the Siegfried Line and thus open a route into Germany.[1] It was a bold and imaginative plan; the operation was certainly a risky undertaking, a gamble even. Some commentators have argued that it was a risk worth taking and that the operation might have succeeded with better fortune, others that it was a vain and foolhardy undertaking.[2]

Market Garden was to be commanded at the operational level by Lieutenant-General Frederick 'Boy' Browning, the Deputy Commander of First Allied Airborne Army (FAAA) and involved two key components.[3] Market was the airborne operation which involved using elements of the newly formed FAAA, commanded by the American Lieutenant-General Lewis Brereton, to seize bridges across canal and river crossings at Eindhoven, Nijmegen and Arnhem in Holland. The bridge at Arnhem sat across the Lower Rhine, the last obstacle into Germany. The bridges around the Eindhoven area were assigned to US 101st Airborne Division under Major-General Maxwell Taylor, with the Nijmegen area assigned to Brigadier-General James Gavin's US 82nd Airborne Division, both of whom had extensive combat experience from Normandy.[4] The final objective, Arnhem, was assigned to British 1st Airborne Division under Major-General Robert 'Roy' Urquhart, supported by the 1st Polish Independent Parachute Brigade under Major-General Stanislaw Sosabowski.[5] These airborne forces were to be carried by transport aircraft from the US IX Troop Carrier Command, commanded by Major-General Paul Williams and elements of 38 and 46 Groups of the Royal Air Force (RAF).[6]

The other element of the operation was Garden. This was the ground assault that involved British Second Army, led by XXX Corps advancing from its bridgehead in Belgium linking up with each airborne division in turn before continuing northward to then outflank the Siegfried Line.

The operation, however, ultimately failed to attain its key objective, to force a secure Rhine crossing at Arnhem and so British forces were unable to penetrate Germany. Although British Field-Marshal Bernard Law Montgomery, General Officer Commanding Twenty-First Army Group and

---

1 Lloyd Clark, *Arnhem: Jumping the Rhine 1944 & 1945* (London: Headline, 2008), 11.
2 Clark, *Arnhem*, 334.
3 TNA, WO 205/313, Operation Orders from Brereton to Browning, 11th September 1944.
4 David Bennett, *A Magnificent Disaster: The Failure of Market Garden, The Arnhem Operation, September 1944* (Newbury: Casemate, 2008), 29.
5 Geoffrey Powell, *The Devil's Birthday: The Bridges to Arnhem 1944* (London: Papermac, 1985), 32.
6 Clark, *Arnhem*, 99-104.

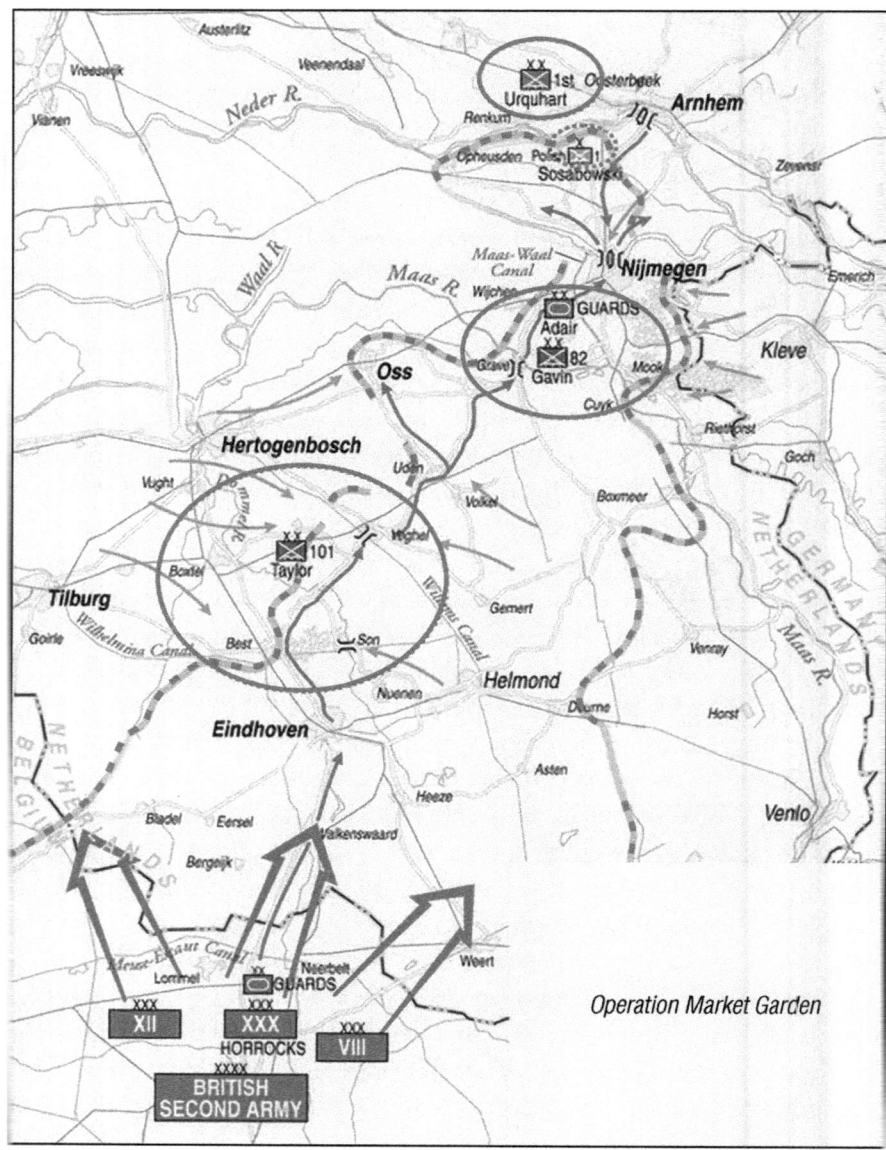

Figure P.1 Operation Market Garden

one of the main if not the prime originator of the plan described it as '90 per cent successful',[7] the operation ultimately failed to capture the Ruhr

---

[7] Cornelius Ryan, *A Bridge Too Far* (London: Hodder & Stoughton, 1974), 532.

nor open the road to Berlin, and the war did not end in 1944. The operation also proved very costly for the Allies, especially the airborne element. Of the units involved, British 1st Airborne Division lost 1,485 men killed and 6,525 captured (over 2,000 of these wounded) out of a total of just under 10,000 men.[8] The US 82nd Airborne Division suffered 3,400 casualties during 56 days of continuous action and US 101st Airborne Division 3,792 casualties (it was not withdrawn from the ground corridor until the end of November).[9]

The operation also sowed greater mistrust and discord amongst senior officers in the Allied ranks.[10] Whilst Montgomery might have described it as 90 per cent successful, in contrast, Prince Bernhardt of the Netherlands said that his country 'could ill afford another Montgomery victory'.[11]

Scope

The operation has been subject to detailed scrutiny over the decades; indeed, a great number of books have been published on the operation, especially British 1st Airborne Division's action at Arnhem. The historical literature typically highlights problems with the plan, often at different levels. For example, John Baynes, biographer of Major-General Roy Urquhart (commander of 1st Airborne Division) suggests that Market Garden was essentially two battles but with no single controlling mind.[12] Peter Harclerode, an Arnhem historian, argues that flaws in both the concept and planning were responsible for the failure of the operation.[13] Another historian, Martin Middlebrook, argues the strategic concept was sound, but the operation suffered from a tactical defeat at Arnhem.[14] Air historian Sebastian Ritchie's comments span different levels. At the tactical level, he identifies the multiple airborne drops at Arnhem as the reason for the failure of 1st Airborne to hold the bridge in sufficient strength, whilst at the operational level, he argues XXX Corps' lack of speed meant that it did

---

8 Martin Middlebrook, *Arnhem 1944: The Airborne Battle* (London: Penguin, 1994), 439.
9 William Buckingham, *Arnhem 1944* (Stroud: Tempus, 2015), 228.
10 James Dickson, *'"But the Germans, General, the Germans, what about them?": The British Assessment of German Fighting Ability and Operation Market Garden, August to September 1944'* (unpublished MA thesis, University of Buckingham, 2017), 128.
11 Buckingham, *Arnhem 1944*, 227.
12 John Baynes, *Urquhart of Arnhem* (London: Brassey's, 1993), 159.
13 Peter Harclerode, *Arnhem: A Tragedy of Errors* (London: Caxton, 1994), 154.
14 Middlebrook, *Arnhem 1944*, 442.

not get to Arnhem in time.[15] Another Arnhem commentator, A.D. Harvey suggests that failures occurred right across the board, at battalion, brigade, division, corps, army and army group levels.[16] There are, therefore, multiple problems at different levels of command; we shall explore all of these.

Operation Market Garden generates conflicting opinions which has sparked much debate and controversy over the years with various participating actors and commentators apportioning blame in different quarters. David Bennett suggests that mistakes were made due to overconfidence and inexperience.[17] Montgomery points to the failure of US General Dwight Eisenhower (the Supreme Allied Commander in Europe) to properly prioritise the operation as the main reason for failure,[18] whereas Bennett points out that Field-Marshal Alan Brooke, the Chief of the Imperial General Staff (CIGS) felt the operation was a lapse in Montgomery's judgement.[19] Ritchie agrees, arguing that the problems with the plan can be traced back to Montgomery and Lieutenant-General Miles Dempsey, commander of British Second Army. Urquhart and Ridgway, both airborne commanders, blamed XXX Corps' slow rate of advance. In reply, Lieutenant-General Brian Horrocks, XXX Corps Commander, blamed the single road along which his formation had to advance.[20] With all the blame flying around both now and then, there was, unsurprisingly, a degree of scapegoating and whitewashing. Montgomery and Churchill both described the operation as successful,[21] the former asserting that the operation had achieved bridgeheads at Nijmegen and liberated parts of Holland.[22] Brereton, as commander of FAAA agreed, arguing Market (the airborne component) was brilliantly successful, but the problem lay with the Second Army advance; his report to his seniors, Generals George Marshall (Chief of the US Army) and Henry 'Hap' Arnold (Commander of the US Air Force) contained a number of half-truths that cover up the reality.[23] The most shameful act of scapegoating, however, was aimed at

---

15 Sebastian Ritchie, *Arnhem: Myth and Reality, Airborne Warfare, Air Power and the Failure of Operation Market Garden* (London: Robert Hale, 2011), 259.
16 A.D. Harvey, Arnhem (London: Cassell, 2001), 193.
17 Bennett, *A Magnificent Disaster*, xii.
18 Field-Marshal Montgomery, *The Memoirs of Field Marshal Montgomery* (London: Pen & Sword, 1958), 296.
19 Bennet-t, *A Magnificent Disaster*, 196.
20 Richard Mead, *General 'Boy': The Life of Lieutenant-General Sir Frederick Browning* (London: Pen & Sword, 2010), 150.
21 Bennett, *A Magnificent Disaster*, xvi.
22 Montgomery, *The Memoirs of Field Marshal Montgomery*, 294.
23 Harclerode, *Arnhem*, 154.

an Allied 'minority group' participant. Montgomery blamed the Poles, accusing them of not wanting to fight, an assertion flying in the face of reality regarding troops whose fellow countrymen in the Polish Home Army were fighting to liberate Warsaw at that time.[24] Browning's own behaviour was more deplorable; he wrote a letter to Brooke stating Sosabowski, the Polish Commander, had been uncooperative, suggesting he be removed from command.[25]

Various commentators, then, have clearly highlighted the errors in the planning and execution of Market Garden, and several potential culprits have been named. None of these commentators have sought to provide a psychological explanation as to why and how these 'culprits' committed their errors. Market Garden therefore provides an excellent case study to develop an understanding of the role of the different psychological factors that underpin the thinking and actions of those involved in planning military operations along with the errors they might make during this process.

Timeframe

This book concentrates on the planning of the operation and therefore mainly focuses on a few weeks at the end of August and the first half of September 1944. The strategic situation prior to the operation, however, had a significant impact on the operation and so the scope of discussion will also touch on events from June 1944 onwards.

Subjects

The book focuses on the airborne element of the plan, Operation Market and so we will concentrate on the senior officers involved in this. At the top of the chain of command, Bernard Montgomery, as commander of Twenty-First Army Group must bear some responsibility. Montgomery devised the operational concept for Market Garden and persuaded General Eisenhower, the Supreme Commander, Allied Expeditionary Force, to back it. The bold nature of the operation and Montgomery's subsequent lack of involvement in its execution was very uncharacteristic, given his previous methodical and controlling approach to operations in the Western Desert and Italy.

---

24 Harvey, *Arnhem*, 196.
25 Bennett, *A Magnificent Disaster*, 238.

This raises the question of why did Montgomery act out of character and how did these factors lead to errors in his planning? We will concentrate on answering this question in part one.

Browning is often portrayed as a self-serving villain of the piece whose personal ambitions overrode his military judgement. As Deputy Commander of FAAA and the officer commanding Market Garden, he did much to ensure the operation went ahead despite mounting concerns. This opens the door to the psychologist's question of what was the motivation that drove and shaped his behaviour and did this lead to errors in his planning? We will address this question in part two.

Major-General 'Roy' Urquhart, commanding British 1st Airborne Division at the sharp end, is often portrayed in the literature as a gallant officer who did his best in difficult circumstances. Critics suggest, however, that he was out of his depth, which led him to make serious errors in his decision-making. This leads us to ask a third question, what were Urquhart's cognitive limitations and how did these lead to errors in his planning? We will address this question in part three.

Methodology

This book uses a case study approach to examine the underlying psychological factors that shaped the thinking of Montgomery, Browning and Urquhart at their respective strategic, operational, and tactical levels. From a psychological perspective, several key questions emerge: what aspects of Montgomery's (difficult) personality explain his (change in) approach; what were the motivational factors that drove and shaped Browning's actions; and what were the pressures placed on Urquhart's decision-making and how did he react to them?

In this book established psychological models and theories are applied to examine these different questions. One outcome from this is that the book draws consistently on secondary sources that describe the individuals' characteristics. For example, Urquhart's autobiography has been referenced regularly because it provides useful insights into his interpretation of events at the time; insights that are integral to the different psychological models used in this work. Clearly, as an autobiography written after the event by a general who led his division to destruction, there is risk of self-justification or distortion of facts. His comments, however, are often supported by other commentators; there is generally a degree of consistency across different observers. This book also draws on John Baynes' biography of Urquhart as it contains descriptions of his personal attributes to populate the different psychological models used. Baynes' biography is sympathetic to its subject, but the descriptions are typically general in nature and are not used in defence of his actions.

Descriptions of each officers' behaviour are also drawn from memoirs, personal letters and contemporary diaries to provide further evidence to support the assessments made about the relevant psychological factors. As discussed above, the use of post facto accounts and memoirs of those involved of course raises issues of source reliability as the writers may simply be seeking to justify their actions post facto or indeed distorting the truth. This book, however, does not use these sources to defend the participants' actions. Instead, these accounts provide background insights into behaviours and characteristics that are useful for the psychological analysis conducted, as these are the characteristics that then have implications for how each officer reacted during the planning of the operation. This approach is further supported where the conclusions drawn from these sources can be triangulated across several accounts, and where these are corroborated from other evidence. The book also accesses various primary sources such as operational orders, situational appreciations and intelligence summaries produced by headquarters staff of each officer. These official records are available, but because of the rushed nature of the planning of Market Garden, there is a lack of detailed papers such as records of meetings and decisions. There is a wealth of material, but this is generally skewed towards the airborne side of the operation. There is also little from the perspective of the air force planners.[26] These are issues to be borne in mind as we work through the thinking behind the planning process.

Structure

This book uses an adaptation of the OODA Loop model to examine each subject as it provides a useful framework to explore their decision-making.[27] 'OODA' is an acronym for a model addressing four stages of the decision-making cycle. The first 'O' is Observation – the assessment of the situation

---

26 Ritchie, *Arnhem*, 12-14.
27 Developments, Concepts & Doctrine Centre, *Land Operations* (Shrivenham, Army Doctrine Publication, AC 71940, 2017), 5-6.

faced by decision-makers. This includes the difficulties faced by individuals in terms of the task-related problems and the situational context in which these existed and served to apply additional pressures. The second 'O' is Orientation – the psychological filters each officer used to make sense of the challenges he faced and judge the impact on his personal goals. 'D' is for Decision – for the purposes of this discussion, the overall strategy or coping mechanism each officer adopted to deal with the situation. This is not his detailed problem-solving activity, but the overall framework or strategy adopted by him within which these processes took place. Finally, 'A' is for Act – the actual problem-solving activities conducted by each decision-maker and in particular, the cognitive biases or errors that he committed. Consistent strands that are explored in each case study are the situational pressures faced by each decision-maker, the coping strategies they adopted to manage these demands, and errors in the planning judgements that they ultimately made. At the Orientation stage, each case study focuses on a different psychological factor that shaped the individual's behaviour: personality (Montgomery), motivational drivers (Browning), and cognitive complexity (Urquhart).

Conclusion

This book differs from the typical historical examinations of Market Garden as it seeks to blend insights from both historical and psychological disciplines into one analysis. It also focuses on the planning of the operation, not the fighting of the battle itself. It hopefully, therefore, contributes to the literature on the subject and in some instances, challenges some accepted viewpoints and opinions. You may be asking yourself, hasn't this all been done before? It would, therefore, be useful to examine the literature briefly to see what the received wisdom is on Market Garden and the extent to which psychological factors are explored.

ACKNOWLEDGEMENTS

I would like to thank Kirstin at Howgate Publishing for taking an interest in this project and kindly agreeing to publish this book. Your encouragement and professionalism have been a great help and made the writing of this book a pleasure. I also owe a huge debt of gratitude to Professor Lloyd Clark whose interest, sage advice and constant encouragement made this book possible.

I would also like to thank my wife, Helen, for putting up with another period of seclusion and further indulgence of my 'obsession with Arnhem'. Full use of the kitchen table has been restored.

Finally, I am most grateful to Dave, Jim and Roger for taking the time to read and comment on the various drafts of this book. They were invaluable in pointing out my historical misunderstandings and correcting my wayward grammar. As ever, any errors or inconsistencies are mine.

Thank you all.

ABBREVIATIONS

ADC	Aide-De-Camp
AHB	Air Historical Branch
BEF	British Expeditionary Force
CIGS	Chief of the Imperial General Staff
DCLI	Duke of Cornwall's Light Infantry
DCM	Decisional Conflict Model
DZ	Drop Zone
FAAA	First Allied Airborne Army
GOC	General Officer Commanding
HQ	Headquarters
LHCMA	Liddell Hart Centre for Military Archives
LZ	Landing Zone
OKW	Ober Kommando West
OODA	Observe, Orientate, Decide, Act
MBTI	Myers Briggs Type Indicator
PR	Photo Reconnaissance
RAF	Royal Air Force
SHAEF	Supreme Headquarters Allied Expeditionary Force
SS	Schutzstaffel
TNA	The National Archives
UK	United Kingdom
UN	United Nations
US	United States
USAF	United States Air Force
WO	War Office

Advance to Contact

Human Factor

John Keegan asserts in *The Mask of Command* that social scientists should not study generalship because of the discipline's tendency to generalise unique characteristics and circumstances to a 'commonality of traits and behaviour'.[1] It is perhaps a rather bold step to take on John Keegan, but I disagree. The demands placed on individuals in specific combat operations means that these situations provide a rich source that can be used to better understand how individuals cope and make decisions in difficult and stressful situations more broadly.[2] Operation Market Garden and the three officers' parts in planning the mission provide an ideal case study that might be used to develop a better understanding of the psychological drivers of those involved in military command. So, what were these human factors? Have they already been addressed in the literature? Let's start on a controversial note, and Norman Dixon.

Norman Dixon

Norman Dixon provides some insight into military decision-making in his controversial book *On the Psychology of Military Incompetence*. He concludes that the planning of Operation Market Garden suffered from specific cognitive biases such as inappropriate risk taking, underestimation of the enemy, and a neglect of unpalatable information.[3] One of his main conclusions is that Field-Marshal Montgomery's desire to beat his rival General George S. Patton, commanding US Third Army, to Berlin overrode all other considerations when he devised the operation. Dixon therefore touches on themes of

---

1 John Keegan, *The Mask of Command: A Study of Generalship* (London: Random House, 1987), 1.
2 Neil Shortland, Laurence Alison & Joseph Moran, *Conflict: How Soldiers Make Impossible Decisions*, (Oxford: Oxford University Press, 2019), 1.
3 Norman Dixon, *On the Psychology of Military Incompetence* (London: Futura, 1976), 145-148.

motivation and cognitive biases, but his treatment of Market Garden is cursory, his use of historical evidence selective in nature and his approach is narrowly psychodynamic. A psychodynamic approach means the analysis suffers from several problems; many of the central concepts are outdated and the analysis tends to neglect broader cultural and social considerations.

These situational factors are discussed at length in the extensive body of historical literature already published on Operation Market Garden, especially 1st Airborne Division's involvement at Arnhem. These works range from memoirs of those involved in the operation to biographies, battlefield guides and more narrative histories by notable historians, Anthony Beevor's Arnhem: *The Battle for The Bridges, 1944*, is an excellent example of this.[4]. Perhaps the most recognised is Cornelius Ryan's *A Bridge Too Far*.[5] Ryan, a war correspondent rather than historian, draws heavily on the personal testimony of those involved in Market Garden to provide a highly detailed account of the operation. Even so, because of Ryan's use of personal testimony and not official records his work lacks academic rigour. Whilst it mainly focuses on telling a narrative history of the operation and does not explore in depth the factors affecting the planners' decision-making, it does provide some useful insights particularly into their character and so will be referred to later in this book.

Academic Papers

Military staff college studies have also been conducted on the subject, mainly at institutions in the UK and the United States. These typically provide critiques of the operation and attempt to identify reasons for its failure to achieve its strategic aim. For example, one thesis written by a USAF Major at the Air Command Staff College discusses Market Garden to explain the principles of war, the author stating that several of these principles were ignored or disregarded in its planning and that this contributed to its failure.[6] Other staff college dissertations focus on various different issues: the failure of commanders to receive, respond to and use intelligence;[7] strategic and operational planners pressing on with the plan despite known risks to test

---
[4] Anthony Beevor, *Arnhem: The Battle for The Bridges, 1994* (London: Penguin, 2018).
[5] Ryan, *A Bridge Too Far*, 121.
[6] William Green, *Operation Market Garden* (Maxwell Air Force Base, Alabama: Defense Technical Information Center, 1984).
[7] Joel Jeffson, *Operation Market Garden: Ultra Intelligence Ignored* (Bolling Air Force Base, Washington DC: Joint Military Intelligence College, 1998).

the concept of mass airborne operations before the war's end;[8] failure in the fields of coordination, integration and synchronisation;[9] the failure of 1st Airborne to achieve surprise and a concentration of force in the early stages of the battle;[10] and finally procedural errors, poor timing and instances of bad luck that were primarily responsible for the breakdown of the 1st Airborne's internal communications.[11] Two papers do refer more specially to individuals. One staff college paper, written by a US Army Reserve Lieutenant-Colonel, does not touch directly on Urquhart's psychological profile but focuses on his lack of airborne experience, the author contending this may have limited the amount he was willing to challenge the air planners over their plan for the delivery of his division and goes on to directly attribute some of the operation's failure to his lack of experience.[12] The other paper, written by a retired American Brigadier-General also focuses on Urquhart's lack of experience commanding airborne operations and again attributes his failure to properly challenge the air planners to this, also highlighting Urquhart was a 'traditional infantry officer' and therefore that he was ill-suited to command an airborne division.[13] This is something we will explore in more detail.

David Houghton has written a useful article that applies the Groupthink model to the operation.[14] He identified several of the antecedent conditions (such as short timescales) required for the phenomenon to occur along with several symptoms of Groupthink (such as stifling dissenters). Overall, Houghton concludes that there is evidence of wishful thinking concerning the planning, implementation and execution of the operation. His focus is on higher levels of seniority within the chain of command (such as Montgomery), he therefore makes only brief mention of Browning and especially Urquhart, noting the latter had intelligence concerns but was unable to push back due to his relatively junior rank and lack of access to Ultra intercepts. Houghton's

---

8 Philip Bradley, *Market Garden: Was Intelligence Responsible for the Failure?* (Maxwell Airforce Base, Alabama: Defense Technical Information Center, 2001).
9 Gordan Van Hook, *Tactical Victory Leading to Strategic Defeat: Historic Examples of Hidden Failures in Operational Art.* (Rhode Island: Naval War College, 1993).
10 Brodie Hoyer, *Operation Market Garden: The Battle for Arnhem* (Maxwell Airforce Base, Alabama: Defense Technical Information Center, 2008).
11 John Greenacre, 'Assessing the Reasons for Failure: 1st British Airborne Division Signal Communications during Operation 'Market Garden'', *Defence Studies*, 4, 3, (2004), 283-308.
12 Elizabeth Coble, *Operation Market Garden: Case Study for Analyzing Senior Leader Responsibilities* (Carlise, PA: United States Army War College, 2009).
13 Michael Clemmesen, 'Combat Case History in Advanced Officer Development: Extracting What is Difficult to Apply', Baltic *Security & Defence Review*, 17, 2, (2014), 34-79.
14 David Houghton, 'Understanding Groupthink: The Case of Operation Market Garden', *Parameters*, 45, 3, (2015), 75-85.

article is useful in that it highlights the utility of the Groupthink model in examining military operations. We will discuss this in more detail in chapter fourteen.

Post-graduate research conducted at academic institutions provides a deeper analysis of some of the key issues, such as the strategic situation that existed at the time,[15] and the Allied (mis)use of the available intelligence,[16] but again these do not explore the situational and psychological factors underpinning decision-making. Taken together, the academic theses and staff college papers discuss the personal qualities of those involved but only to a limited extent. A clear sense of the decision-makers' personal qualities as individuals and as military commanders does not emerge from a reading of these works.

Memoirs

Unsurprisingly, our three protagonists feature in the published historical literature dealing with both Market Garden in general and the battle at Arnhem more specifically. The main focus in this literature is on a number of key themes: the strategic situation in Northwest Europe in September 1944;[17] the personal rivalries existing at senior Allied levels;[18] the nine days duration of the operation itself;[19] the experiences of those who fought in the battle,[20] including from the German perspective;[21] and an analysis of the causes for the failure of the operation.[22] What is again lacking in these and other works is an analysis of the underlying psychological factors that led to the key errors being made; the focus is typically on what went wrong rather than how or why.

Roy Urquhart himself wrote a memoir in 1958, titled *Arnhem*, which proves useful as it provides some insight into his earlier career and his thinking about the operation itself;[23] referring to the difficulties and constraints he

---

[15] Roger Cirillo, 'The Market Garden Campaign: Allied Operational Command in Northwest Europe,1944' (unpublished doctoral thesis, Cranfield University College of Defence Technology, 2002), 7-50.
[16] Dickson, '"But the Germans, General, the Germans, what about them?"', 116.
[17] Richard Holmes, *Battlefields of the Second World War* (London: BBC, 2001), 134-136.
[18] David Irving, *The War Between the Generals* (London: Penguin Books, 1981), 267-288.
[19] John Nicholl & Tony Rennel, *Arnhem: The Battle for Survival* (London: Penguin, 2011), 1-17
[20] Middlebrook, *Arnhem 1944*, 59-74.
[21] Robert Kershaw, *It Never Snows in September* (London: Ian Allan, 1990), 385.
[22] Beevor, *Arnhem*, 10.
[23] Roy Urquhart, *Arnhem* (London: Pen & Sword, 1958), 1-31.

faced when planning the operation. Clearly, Urquhart may have written his memoir to defend himself or justify his actions and so, as already highlighted, caution must be exercised in its use as it is important to consider the aims and motives of individuals involved. John Baynes (*Urquhart of Arnhem*) has published a useful biography of Urquhart that sets out his early life and career in the military.[24] Victor Dover (*The Sky Generals*) also provides useful insights into the thoughts and actions of the key decision-makers and includes several quotes from interviews he conducted with Urquhart.[25]

Other participants in the battle have also published memoirs, including Lieutenant-Colonel John Frost (*A Drop Too Many*),[26] and Stanislaw Sosabowski (*Freely I Served*).[27] As with Urquhart, their motives for writing their memoirs should be borne in mind, but these works do provide useful firsthand comments about the operation, providing some insight into Urquhart's psychological profile but understandably do not tie these together into a clear framework through which his decision-making can be viewed and understood. Other key figures involved in Market Garden have also produced memoirs or had biographies written about them. Richard Mead produced a biography, *General 'Boy': The Life of Lieutenant-General Sir Frederick Browning*, but (as Browning's papers were destroyed), this work lacks substantive source material.[28] The other key figure in Market Garden, Field-Marshal Montgomery published his own memoir where he focuses on the broader strategic and operational issues.[29] Nigel Hamilton has also published a biography of Montgomery in three volumes; the first covers his early years,[30] the second his war service up to 1942,[31] and the third, the rest of his life.[32] These works provide valuable insights into his personality and this book makes extensive use of these.

---

[24] Baynes, *Urquhart of Arnhem*, 3-49.
[25] Victor Dover, *The Sky Generals* (London: Cassell, 1981), 120-154.
[26] John Frost, *A Drop Too Many* (London: Pen & Sword, 1994), 203-232.
[27] Stanislaw Sosabowski, *Freely I Served: The Memoir of the Commander – 1st Polish Independent Parachute Brigade 1941-1944* (London: Pen & Sword, 2013), 138-191.
[28] Mead, *General 'Boy'*.
[29] Montgomery, *The Memoirs of Field Marshal Montgomery*, 58.
[30] Nigel Hamilton, *The Full Monty: Montgomery of Alamein 1887-1942* (London: Penguin, 2001).
[31] Nigel Hamilton, *Monty: Master of the Battlefield 1942-1944* (London: Hamish Hamilton, 1983).
[32] Nigel Hamilton, *Monty: The Field Marshal 1944-1976* (London: Hamish Hamilton, 1986).

Historical Literature

The difficulties and constraints faced by the decision-makers are discussed at length in several published military history works. Peter Harclerode in *Arnhem: A Tragedy of Errors*, highlights the fundamental problems with the plan for Operation Market Garden.[33] David Bennett's *A Magnificent Disaster* neatly introduces several themes that we will examine further from a psychological perspective, for example, the problems that existed with 1st Airborne Division's training.[34] William Buckingham's *Arnhem 1944* also highlights the lack of training and battle experience in 1st Airborne Division, the assumptions that were made about the battle readiness of the Wehrmacht and the underestimation of its ability to react to the airborne landings.[35] Martin Middlebrook (*Arnhem 1944*) lists the following problems: over optimism about the proven German powers of recovery; problems with the air plan; failure to prioritise the capture of Nijmegen bridge (in the middle of the corridor); and a lack of drive by XXX Corps, tasked to relieve the airborne divisions.[36] Another theme explored at length in the literature is the Allied air plan that constrained Urquhart's planning for the operation.[37] Buckingham comments on the insistence on a single air lift, versus two on the first day, no *coup de main* operations to seize the bridges, and distances from landing and drop zones to the objectives.[38] At a tactical level, Christopher Hibbert (*The Battle of Arnhem*) points out that 1st Airborne's advance into Arnhem was too cautious.[39] We will examine all of these issues in detail later.

Other factors likely to have affected the mindset of our three subjects are discussed in a number of articles in a book edited by John Buckley and Peter Preston-Hough (*Operation Market Garden: The Campaign for the Low Countries, Autumn 1944: Seventy Years On*).[40] This is a collection of works that discuss various issues including: the impact of the manpower shortage that existed at the time;[41] the stiffer German resistance encountered

---

33 Harclerode, *Arnhem*, 46-63.
34 Bennett, *A Magnificent Disaster*, 1-42.
35 Buckingham, *Arnhem 1944*, 37-52.
36 Middlebrook, *Arnhem 1944*, 442-44.
37 Ritchie, *Arnhem*, 249-261.
38 Buckingham, *Arnhem 1944*, 231-234.
39 Christopher Hibbert, *The Battle of Arnhem* (London: Batsford, 1962), 206.
40 John Buckley & Peter Preston-Hough, *Operation Market Garden: The Campaign for the Low Countries, Autumn 1944: Seventy Years On* (Solihull: Helion, 2016), 15-18.
41 John Peaty, 'Operation Market Garden: The Manpower Factor', in *Operation Market Garden: The Campaign for the Low Countries, Autumn 1944: Seventy Years On*, eds. John Buckley & Peter Preston-Hough, (Solihull: Helion, 2016), 58-73.

because of the week's delay in mounting the operation;[42] and ignoring the lessons learned from previous airborne operations.[43] A failure to learn (as well as to anticipate and adapt) is highlighted by Eliot Cohen and John Gooch in *Military Misfortunes: The Anatomy of Failure in War*, as major reason(s) why military organisations fail.[44] Although not specifically about Arnhem, this goes some way towards providing some insight into the underlying psychological reasons for the problems encountered in the conception and planning of Market Garden. It does, therefore, begin to touch on some of the issues that this book seeks to explore but does not deal in depth with the underlying psychological attributes examined, such as a commander's cognitive capacity and motivational drivers, and possible impact these have on planning and decision-making.

In summary, nearly all the discussions in the Market Garden literature are devoted to the prevailing circumstances existing at the time the planning for the operation was conducted, along with the various flaws or errors made in the conception of the operation. The motives that underscore the intentions and actions of the central characters in the situation are discussed as factors but are often simply described as personality clashes or individual jealousies. More importantly, no attention is given as to how these factors affected the mindsets of those involved. Montgomery, Browning and Urquhart are, of course, not unique as military commanders in being required to make difficult decisions in stressful situations with many uncertainties, indeed, this is what Carl von Clausewitz referred to as the 'friction of war'.[45] Although not unique in facing these difficulties, Market Garden does provide a good case study for examining the psychological aspect of the friction of war because each decision-maker had to deal with both external pressures and internal issues. As part of this process, each did commit different errors in decision-making and suffered from various cognitive biases. This book attempts to explore these issues by raising several key questions.

---

42 Jack Didden, 'A Week Too Late?' in *Operation Market Garden: The Campaign for the Low Countries, Autumn 1944: Seventy Years On*, eds. John Buckley & Peter Preston-Hough, (Solihull: Helion, 2016), 74-98.
43 Sebastian Ritchie, *Learning to Lose? Airborne Lessons and the Failure of Operation Market Garden* in *Operation Market Garden: The Campaign for the Low Countries, Autumn 1944: Seventy Years On*, eds John Buckley & Peter Preston-Hough, (Solihull: Helion, 2016), 19-36.
44 Eliot Cohen & John Gooch, *Military Misfortunes: The Anatomy of Failure in War* (New York: Free Press, 1990), 26-28.
45 Carl von Clausewitz, *On War* (Princeton: Princeton University Press, 1984), 119-121.

Questions

In conclusion, the following discussion explores the main question that this book addresses: what were the different psychological factors that affected Montgomery, Browning and Urquhart's roles in the planning of Operation Market Garden in the weeks leading to September 1944? This primary question, using the OODA Loop model as a framework, is broken down into four supporting subsidiary questions: what were the external pressures perceived by the decision-makers (Observation); how did personal factors (Montgomery's personality, Browning's motivation and Urquhart's cognitive capacity) affect the way they interpreted the situation and the conclusions they reached (Orientation); how did the ways in which they attempted to cope with the situation affect their planning (Decision); and finally what were the errors (cognitive biases and judgement heuristics) that were manifested in the planning undertaken by them (Action)? Let's begin at the top, with Montgomery and put ourselves into his shoes as we explore his thinking at the first stage of the OODA Loop – Observation.

Part One

Reproduced with kind permission of the Imperial War Museum.

Montgomery

1

Montgomery's Observation – Discord

'One powerful full-blooded thrust across the Rhine'.[1]

Stalemate

Montgomery's Situation

We will begin our exploration of Montgomery's decision-making by looking at the strategic situation as of 15th September; lets join him at his Tactical Headquarters in Belgium as he is taking stock of the situation that faced the Allies in the European Theatre of Operations. British and American forces had finally broken out of the Normandy bridgehead and had advanced across Northwest Europe. Twenty-First Army Group had moved rapidly along the coast in Northern France, crossed into Belgium, liberating Brussels on 3rd September. To his right, was the US Twelfth Army Group under General Omar Bradley. Further south, the US Sixth Army Group under Lieutenant Jacob Devers was advancing northwards having landed on the French Mediterranean coast in August.[2] This advance was in accordance with General Eisenhower's strategy of a broad front approach to weaken the main German defensive fronts and prevent consolidation.[3]

The progress, after the hard slog of the Normandy bridgehead, had been rapid. The British XXX Corps, under his command as part of Second Army,

---

1 Hibbert, *The Battle of Arnhem*, 11.
2 Clark, *Arnhem*, 1-31.
3 Roger Cirillo, 'Market Garden and the Strategy of the Northwest Europe Campaign' in *Operation Market Garden: The Campaign for the Low Countries, Autumn 1944: Seventy Years On*, eds. John Buckley & Peter Preston-Hough, (Solihull: Helion, 2016), 42.

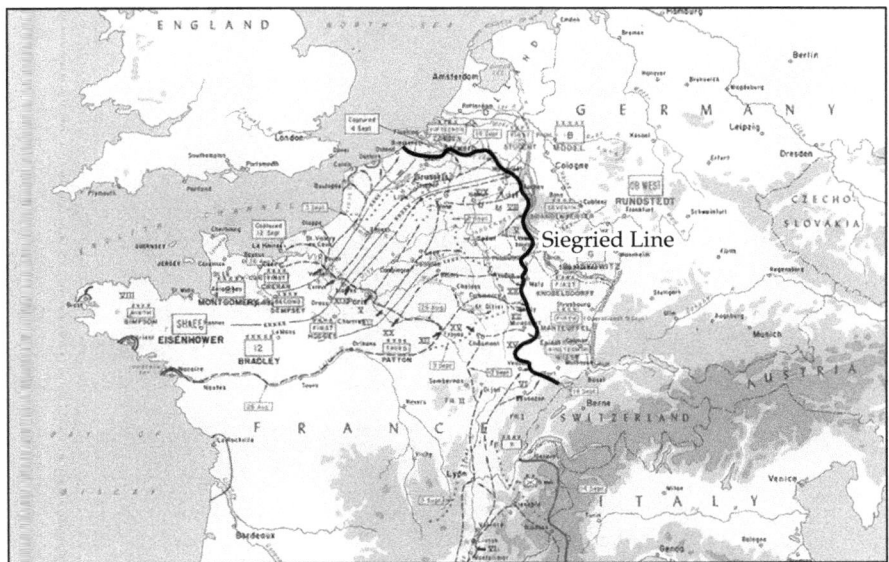

Figure 1.1 By September 1944, the Allies had pushed right up to the Siegfried Line

for example, had advanced at an average of 50 miles a day during August and early September.[4] Second Army, by 10th September, as can be seen in Figure 1.1, had established bridgeheads over the Albert Canal in Belgium and liberated Brussels, capturing the port of Antwerp, although crucially not the Scheldt Estuary which prevented the opening of the port to receive supplies.[5]

The rapidity of the August advance had led to a feeling of confidence that the German Wehrmacht was as good as beaten. A recent intelligence summary from Eisenhower's Supreme Headquarters Allied Expeditionary Forces (SHAEF), dated 4th September summarised that view nicely, stating that the Wehrmacht was 'no longer a cohesive force but a number of fugitive battle groups, disorganised and even demoralised, short of equipment and arms'.[6] Another summary, from Second Army dated 5th September, seemingly suggested the war would soon be over: 'The August battles have done it; brought the end of the war in Europe in sight, almost within reach'.[7]

---

[4] Powell, *The Devil's Birthday*, 17.
[5] Irving, *The War Between the Generals*, 269.
[6] Powell, *The Devil's Birthday*, 16.
[7] LHCMA, Dempsey Papers, British 2nd Army Daily Intelligence Summary, No. 93, 5 September 1944.

Victory Euphoria

Looking with hindsight, we can detect, in early September, a sense of confidence, indeed, overconfidence amongst the Allies that the war was won, since described as a kind of 'victory euphoria'.[8] The problem was that this victory euphoria created a strong sense of optimism that dulled thinking and caused problems in the way that decision-makers viewed the situation.[9] Blinded by an eagerness to end the war, they would mould the plan for Market Garden to fit what was an essentially flawed premise that the Wehrmacht was beaten, and one final push should end the war.[10] This belief was flawed for several reasons, one of these was logistics; Montgomery and the Allies had a problem.

Montgomery's Supply Problem

Montgomery was frustrated at the acute supply problem. The major ports on the French coast had not been opened and supplies were still being unloaded onto the Normandy beaches then driven by truck over 300 miles to the frontline.[11] This was causing severe problems for his frontline units. He had signalled Eisenhower on 7th September informing his boss that over the last two days, he had received only 375 of the 1,000 tonnes that he needed per day. The leading elements of British Second Army, the Guards Armoured and 11th Armoured Divisions, were halted at Antwerp and Brussels respectively.[12] His American allies faced a similar problem, Patton, his great rival, had earlier been forced to halt his advance at the Meuse River.[13] The issue: there were simply not enough supplies flowing to maintain all the Allied forces. Priority over and access to supplies was a hotly contested issue and was a key factor in the debate that had been raging about which strategy to adopt going forward, without opening a major channel port, it was unlikely to be resolved soon. The need to clear the Scheldt Estuary to Antwerp loomed large in both his and Eisenhower's deliberations. Supplies were not the only thing Montgomery was short of; Britain was also running out of men.

---

8 Harclerode, *Arnhem*, 157.
9 Bennett, *A Magnificent Disaster*, 7.
10 Ibid., xvi.
11 Ibid., 219.
12 Clark, *Arnhem*, 10.
13 Harclerode, *Arnhem*, 22.

Montgomery's Manpower Shortage

By September 1944, there was a severe manpower problem in the British Army in general and more specifically, in Montgomery's own Twenty-First Army Group. By the end of the month, Twenty-First Army Group would be short of 10,000 infantrymen and nearly 15,000 by October. Twenty-First Army Group had, since June, sustained 70,000 casualties of which over half were infantry; his infantry divisions had suffered a 40 per cent casualty rate. This meant his infantry battalions were chronically under-strength. Since June he had been forced to break up one infantry division and one brigade to bring other units up to strength. Other measures had been taken to redress this shortfall, RAF and Navy personnel were being transferred to the Army and Royal Artillery gunners were being re-trained in the infantry role. This crisis meant that his spearhead, the infantry, was only 28,800 men strong.[14] The issue added impetus to the need to bring the war to an end. Montgomery would later write in his memoirs: 'The British economy and manpower situation demanded victory in 1944', and as we shall see, 'my plan [Market Garden] offered the only possibility of bringing the war to a quick end'.[15] The lack of supplies and men were not the only reasons for Montgomery wanting to act quickly, German resistance was beginning to stiffen.

Stiffening Resistance

It was clear to Montgomery that over the past few weeks the Wehrmacht had regained some of its cohesion, especially the units facing Miles Dempsey's Second Army. The German High Command, OKW, had been frantically scraping together new divisions from 'paratroops and pilots, policemen and sailors, boys of 16 and men with duodenal ulcers', but however scratch these units might have been, they were beginning to put up a stiff resistance that had slowed his advance.[16] Prominent in this strengthened resistance were Luftwaffe General Kurt Student's newly formed First Parachute Army and a hastily formed Kampfgruppe under General Kurt Chill. The latter formation would later prove to be particularly problematic slowing up the British advance during Market Garden.[17] The Wehrmacht's regeneration had

---
14 TNA, WO 32/10899, Directorate of Staff Duties: Manpower.
15 Harclerode, *Arnhem*, 159.
16 Hibbert, *The Battle of Arnhem*, 17-19.
17 Didden, *A Week Too Late?*, 95-97.

become evident when Second Army renewed its attack on 6th September; the appreciation of this fact had led Montgomery to cancel Comet, (the airborne operation that was the predecessor to Market Garden), the operation that had been central to his future plans.[18]

As Montgomery examined the situation, it was clear that the Allied advance had slowed down. The supplies deficit had stalled units across the Allied front. It was also clear that after their headlong retreat, the Wehrmacht was regrouping and offering stiffer resistance. The sense of victory euphoria still clung on doggedly, but the situation posed a conundrum for both him and the senior officers at SHAEF: what was the best strategy to deal with the situation?

Options

Montgomery's Assessment

As Montgomery assessed the possible ways forward, it was clear that there were essentially four options for what to do next. The first was to pause the advance and concentrate on improving the supply situation. The best way to do this was to open the channel ports and Antwerp especially. The Germans would need to be cleared from the banks of the Scheldt Estuary and the island Walcheren.[19] The advance could then continue once the supply situation had been improved. It had been suggested that the Allies' strategic reserve, First Allied Airborne Army (FAAA), could be dropped on Walcheren to seize it. The planners at FAAA had rejected the proposal as the ground was unsuitable for airborne operations.[20] Without this bold move, Montgomery knew that the capture of the island and the clearing of the estuary would be a deliberate task and would take some time to complete. The problem with this option, in his view, was its cost in gifting the Wehrmacht further time to regroup and prepare a stronger defence, particularly along the Siegfried Line, the pre-war defensive positions on Germany's western border. The other three options available would all keep up the pressure on the Wehrmacht by maintaining the Allied advance; this he knew, was important.

The second option was to maintain the momentum of the advance and keep up the pressure by maintaining a steady forward movement

---

18 Clark, *Arnhem*, 24.
19 Clark, *Arnhem*, 96.
20 Hamilton, *The Full Monty*, 451.

across a broad front. This strategy had been proposed by Eisenhower, one that he championed when he took over from Montgomery as Land Forces Commander on 1st September.[21] The term 'broad front' was in fact a misnomer as the strategy would involve two separate concentrated thrusts along two diverging axes of advance, one to the south and one to the north of the Ardennes forest.[22] Montgomery was vehemently opposed to this strategy, its main issue as far as he could see, was the logistics problem it would create since both the Allied army thrusts would need to be supplied. This would mean the rate of advance would be slow, thus giving the Wehrmacht more time to reorganise.

The other two options meant prioritising either one of the narrow thrusts. The southerly thrust would involve an advance through Aachen, aimed at seizing one of Germany's industrial centres, the Saar. It would be a wholly American affair, involving General Omar Bradley's Twelfth Army Group, consisting of US First, Third and Ninth Armies. The problem with this approach, (to be spearheaded by Patton), was the number of fortifications in the area, including the Siegfried Line and that it would take away the logistics assigned to his own northern thrust.[23]

Montgomery's northern thrust proposal was to bypass the Siegfried Line by outflanking it to the north. He had written to Eisenhower on 4th September, claiming that 'One powerful full-blooded thrust across the Rhine and into the heart of Germany, backed by the whole of the resources of the Allied Armies, would be likely to achieve decisive results'.[24] This was essentially a reverse of the 1914 Schlieffen Plan that the Germans had re-used in May 1940 in their own assault across the Meuse in the opposite direction. Once he had got a bridgehead across the Rhine and outflanked the Siegfried Line, Montgomery believed he could capture the Ruhr industrial area in northern Germany. This, he calculated would end the war in three months as the Ruhr was Germany's major area for extraction of coal and production of steel.[25] The northern thrust also provided a more direct route to the ultimate prize, Berlin and involved shorter lines of communication as the advance kept it closer to the channel ports.[26] He also favoured the northern thrust because the North German plains provided better terrain for armour than

---

21 Bennett, *A Magnificent Disaster*, xv.
22 Buckingham, *Arnhem 1944*, 70.
23 Buckingham, *Arnhem 1944*, 71.
24 Hibbert, *The Battle of Arnhem*, 11.
25 Ibid., 22.
26 Powell, *The Devil's Birthday*, 23.

the forested hills to the south. A successful attack in the north would have the added political advantage of ensuring Britain a prominent role in the politics of a postwar Europe, (it would also serve Montgomery's growing ego – more of this in the next chapter).[27] The difficulty, he had to admit, was the multiple rivers and waterways in Holland that would need to be crossed.[28] But this problem could be solved by using FAAA to secure the bridges over these obstacles.

Montgomery had two options for executing the northern thrust, or more correctly two possible axes of advance, a more southerly route through Wesel or a more northerly one through Arnhem. Wesel, with its single crossing point over the Rhine was Miles Dempsey, the commander of British Second Army's preferred option, he had advocated its selection at a meeting with Montgomery on 10th September.[29] The route offered a shorter distance into Germany and meant the British thrust would not diverge from the US First Army to its south, who would, therefore, be able to offer it support. (Brigadier David Belchem, who was deputising for Freddie de Guingand as Montgomery's Chief of Staff at Twenty-First Army Group, would later state that German generals, with whom he spoke postwar, claimed that their forces in the area were 'virtually non-existent' and that the British soldiers could have simply 'bicycled along the road to Wesel' virtually unopposed).[30] The main problem with Wesel was that any airborne operation there would expose the transport aircraft to the flak concentrations in the Ruhr, something the air planners were not keen on.[31] The other option was to advance through the Arnhem area.

Attacking through Arnhem, in Montgomery's estimation, had several advantages, one was the avoidance of the German fighter gruppens and flak concentrations in the Ruhr. Routing to the north would also avoid the more northerly defences of the Siegfried Line.[32] (There was a third consideration, that would emerge as the situation unfolded, but more of that later). The second advantage was selection of Arnhem as the final objective meant the plan for a recently devised but cancelled airborne operation in the area, Comet, could be recycled.[33] Arnhem as Montgomery also knew, had its disadvantages. As

---

27 Clark, *Arnhem*, 1.
28 Harclerode, *Arnhem*, 17.
29 TNA, WO 285/9, Dempsey Diary, 9 September 1944.
30 Powell, *The Devil's Birthday*, 26.
31 Hibbert, *The Battle of Arnhem*, 22.
32 Powell, *The Devil's Birthday*, 25.
33 Ritchie, *Arnhem*, 114.

Dutch army staff appreciations had concluded, the area through which the advance would run was easily defendable and a subsequent breakout after a Rhine crossing would be difficult.[34] Arnhem, because of its more northerly route would expose the advancing troops' eastern flank as they broke away from the Americans to the south.[35] Whichever route was chosen, he would need a concentration of force and so he had argued for a powerful punch of 40 divisions.[36] For this, he would need to take US divisions under his command and take all of the limited supplies, effectively halting the American advance in the south. This was the third problem he now faced, one that he knew from bitter experience, and he was ill-equipped to deal with: politics.

Politics

The situation in North-west Europe in late summer 1944 was fraught politically. Eisenhower, as Supreme Commander, was in a difficult position. His main problem was trying to keep the alliance together. A key issue for Eisenhower was the need to consider public opinion in America, which would not take kindly in an election year, to priority being given to the British when the US was now the senior partner in the alliance.[37] Whilst he favoured a broad front approach, he had his two direct subordinates championing competing initiatives, Bradley in the south and Montgomery in the north, with each general believing that he was favouring the other. A growing resentment was being felt by the American generals in the theatre of operation, with Patton the biggest protagonist. He disliked the British and seemed to view Montgomery as 'the real opponent'.[38] Although too much has been made of this over the years, a number of commentators would agree that a feud had existed between the two since the Sicilian campaign. Harclerode goes so far as to suggest that it was the 'discord and disagreement [that] had festered continually between Montgomery and Generals Omar Bradley and George Patton' that sowed the 'seeds of failure' for Market Garden.[39]

The problem was compounded by the fact that Montgomery could not grasp the political imperative of the American contingent needing to be (seen to be) in charge. Montgomery's ignorance of this fact and persistent advocacy

---

[34] Ibid., 107.
[35] Ibid., 256.
[36] Montgomery, *The Memoirs of Field Marshal Montgomery*, 266.
[37] Irving, *The War Between the Generals*, 213.
[38] Irving, *The War Between the Generals*, 15.
[39] Harclerode, *Arnhem*, 155.

for his northern thrust would help fuel the rancorous debate that would rage back and forth about which option to adopt to force back the Wehrmacht into Germany, an argument that, as we shall see in the next chapter, would have an unfortunate effect on Montgomery.[40] It will be worth following it as it developed, as it helps to highlight the twists and turns of the debate and Montgomery's growing frustration. Let's return to Montgomery's caravan and put ourselves back in his position.

Montgomery's Frustration

Montgomery's memoirs clearly describe the protracted debate that had been causing him so much frustration.[41] The argument had started in earnest in mid-August when he had met with Bradley on the 17th of that month to discuss the strategic situation. Montgomery had suggested a strong thrust in his northern sector with Twenty-First Army Group advancing on the western flank through southern Holland, and Bradley's Twelfth US Army Group moving forward on the eastern flank through Aachen. Bradley, at this point in the proceedings seemed to agree. So far, so good. The problem was Eisenhower disagreed. Three days later, at a conference at his headquarters in Granville on the Normandy coast, Eisenhower not only announced his intention to take over as Land Forces Commander on 1st September but also his intention to 'push forward on a broad front with priority on the left' to relieve the supply problem and then pivot to the east to 'directly threaten the Ruhr'. This was fine and suited Montgomery, but crucially, Eisenhower also wanted Bradley's Army Group to head towards the Saar region.[42] Montgomery had not agreed with this approach and had sent Freddie de Guingand, his Chief of Staff to see Eisenhower with some notes emphasising the need to advance northwards as a cohesive force. Despite de Guingand's best efforts over a two-hour meeting, he had failed, and Eisenhower had stood his ground. Undeterred, Montgomery had tried again.

Montgomery had written to Eisenhower on 22nd August to again promote his case. He argued that the 'quickest way to win this war is for the great mass of the Allied Armies to advance northwards [and that] single control and direction of the land operations is vital for success. This is a

---

40 Cirillo, *Market Garden and the Strategy of the Northwest Europe Campaign*, 56.
41 Montgomery, *The Memoirs of Field Marshal Montgomery*, 267.
42 Harclerode, *Arnhem*, 14.

WHOLE TIME job for one man',[43] (him obviously, not Eisenhower). This correspondence had not had the desired effect, and so Montgomery had invited Eisenhower to meet him at his tactical headquarters the next day. Montgomery first flew the next morning to meet Bradley on 23rd August, before meeting with Eisenhower in the afternoon. To his surprise, Bradley told him that he had changed his mind and now supported the broad front strategy with his Army Group advancing towards Metz and the Saar region. When Montgomery met with Eisenhower, he had urged him to make a firm decision about the direction of advance, whilst criticising the broad front strategy, stating it would run out of momentum due to supply problems. He again pushed his northern thrust as a better alternative and asked for 12 American divisions to take part in the operation.[44] Eisenhower had rejected the idea on the basis that it was too risky to commit the limited available resources to one venture and that American public opinion would not stand US troops serving under a British general. It would also involve halting Patton's advance in the south, which was again, politically unacceptable. Montgomery had then raised the subject of command, stating that as Supreme Commander, Eisenhower 'must sit on a very lofty perch in order to be able to take a detached view of the whole intricate problem … Someone must run the land battle for him'.[45] He had even offered to serve under Bradley if American public opinion was an issue. Eisenhower refused again but did offer the temporary assistance of US First Army and, more importantly FAAA with the suggestion that it be used in a bold and decisive fashion. Montgomery's hopes were raised by these concessions but dashed ten days later when Eisenhower gave Patton approval to continue his advance across the Moselle. The same problem had returned, by continuing with two advances, there would not be enough supplies for his northern thrust.[46] He was back to square one.

Montgomery had suffered a further setback when he attended a meeting on 3rd September where Bradley, taking advantage of a temporarily incapacitated Eisenhower, announced his intention to limit his support to Twenty-First Army Group and instead put his full weight behind Patton's advance.

Prompted into action, Montgomery had then ordered Dempsey to advance towards Wesel and Arnhem on 6th September. This route had

---

[43] Montgomery, *The Memoirs of Field Marshal Montgomery*, 267-268.
[44] Harclerode, *Arnhem*, 14.
[45] Clark, *Arnhem*, 19-20.
[46] Ritchie, *Arnhem*, 93.

previously been rejected because of the need to cross a number of water obstacles, but this could be solved by the use of FAAA to secure bridges over the Rhine between Wesel and Arnhem; this had been Operation Comet.[47] He had switched the final objective to Arnhem (probably as a result of a meeting with Browning on 3rd September) after the RAF protested about the threat posed by German flak concentrations in the Ruhr.[48]

Montgomery had then gone into bat again, writing a long message to Eisenhower on 4th September, asking to be given the priority for resources. He reiterated several points including the need for one 'full-blooded thrust', that this should have all the resources devoted to it and that, in his opinion, the northern thrust was the better option. The message had exasperated Eisenhower due to its arrogant tone;[49] Montgomery had written that 'if you are coming this way perhaps you could look in and discuss it ... Do not feel I can leave this battle just at present'. The problem was exacerbated by the message crossing with one from Eisenhower in which he re-emphasised his intention to proceed with two thrusts.[50] Eisenhower had replied the next day, on 5th September, but the communications from his headquarters was so poor that Montgomery received the second half of the message on 7th September and the first part on 9th September. Eisenhower stated in his message that he agreed with the idea of a strong thrust to Berlin but that this could not be supported at the expense of other operations, reiterating his intention to maintain two thrusts as part of the broad front strategy.[51] He also stated, in the second half of the message (which arrived first on 7th September) that 'I have always given and still give priority to the Ruhr ... and the northern route of advance'.[52] Montgomery had then replied on 7th September, emphasising the issue with logistics and asking Eisenhower to come and see him.[53] He had also been emboldened enough by Eisenhower's comment about priority to consider expanding Comet into a much larger operation – Market Garden, (the objective of which would be changed as a result of a message received by Montgomery two days later, or so the story goes).[54]

---

47 Clark, *Arnhem*, 19.
48 Ritchie, *Arnhem*, 96.
49 Ryan, *A Bridge Too Far*, 79.
50 Montgomery, *The Memoirs of Field Marshal Montgomery*, 271-272.
51 Harclerode, *Arnhem*, 24.
52 Montgomery, *The Memoirs of Field Marshal Montgomery*, 272-273.
53 Harclerode, *Arnhem*, 25.
54 Montgomery, *The Memoirs of Field Marshal Montgomery*, 274.

V2 Impact

Montgomery states in his memoirs that he received a message on 9th September from the War Office in London informing him that a new terror weapon, the V2 rocket had started to fall on London on the 8th and urging him to take immediate action to clear the launch sites in Holland. This, he claims was the deciding factor in convincing him to switch the line of advance away from Wesel and towards Arnhem.[55] The problem with this account is that Arnhem had already been set as the objective for Comet and so the V2 message was probably just an excuse.[56] The advantage of switching to Arnhem was that it would take the axis of advance on a more divergent route from the Americans in the south placing it more firmly under Montgomery's aegis. Regardless of when the decision was made, the outcome was that the northern thrust was now, in Montgomery's mind, a larger operation aimed at Arnhem; this was what was discussed when he finally met with Eisenhower at Brussels Airport on 10th September. Time to rejoin Montgomery in his battle with Eisenhower.

Montgomery's Meeting

The meeting had not started well. Montgomery had launched into a vitriolic critique of Eisenhower's strategy, to the point where the latter had stopped him and reminded him that he was his boss.[57] Montgomery, chastened by the rebuke went on to describe his plan for Market Garden. Eisenhower had been taken by the sort of imaginative use of FAAA that his superiors, Generals Marshall and Arnold had been urging, and he had been pressing Montgomery to do, but he was unmoved, telling Montgomery that he was nuts' to think he could manage a single thrust to Berlin.[58] Montgomery tried again, this time playing his trump card, the V2 message from the War Office. Finally, Eisenhower had given approval to the operation because he wanted a bridgehead over the Rhine, but did not agree to the further exploitation after Arnhem due to the logistical constraints.[59] Operation Market Garden was therefore on, but there were a few more twists and turns.

---

55 Ibid., 274.
56 John Buckley, *Monty's Men: The British Army and the Liberation of Europe* (London: Yale University Press, 2014), 212.
57 Harclerode, *Arnhem*, 27.
58 Ryan, *A Bridge Too Far*, 91.
59 Ritchie, *Arnhem*, 114.

Montgomery signalled to Eisenhower the next day to say that due to the supply problem he could not launch the operation until 23rd or 26th September.[60] Eisenhower had sent his Chief of Staff, US General Walter Bedell Smith to meet with Montgomery on 12th September, and when Bedell Smith promised to accelerate the provision of the bulk of the required supplies, Montgomery had set the date of the operation as 17th September, now only five days away.[61] Montgomery had been delighted with what he saw as a great victory but he had been almost immediately disappointed when he subsequently received a message on 13th September informing him he would not get all he had been promised.[62] The final moves in the game had come on 14th September when Eisenhower announced that he agreed to support the advance that Patton had already started in defiance of the direction he had been given to stop. This move practically ensured that Montgomery would not get the supplies he needed. On 15th September, Eisenhower then sent out a directive to all his senior commanders in which he set out not one but five objectives, seizure of the Ruhr, Saar and the three main channel ports.[63] This had effectively put paid to Montgomery's hopes of a concentrated fully supported northern thrust.

Montgomery was, with only two days to go, exhausted by the long and frustrating argument. He was clearly faced with a difficult problem which was made worse by the intra-Allied politics and relationships involved. The task pressure he faced was acute. Added to these issues were pressures induced by the broader situational context.

Situational Pressures

Influence Factors

Actions can be strongly influenced by different contextual or social factors that exist beyond the pressures induced by the task itself. These social influence factors have been studied by an American Psychologist, Robert Cialdini. Cialdini spent three years working in fund-raising organisations, second-hand car dealerships and telemarketing companies to investigate the sorts of ploys or approaches used by sales staff and marketing professionals

---

50 Irving, *The War Between the Generals*, 277.
51 Powell, *The Devil's Birthday*, 27.
52 Harclerode, *Arnhem*, 29.
53 Montgomery, *The Memoirs of Field Marshal Montgomery*, 277.

to influence customer choices in everyday life. Although these are specific techniques used in sales and marketing contexts, they work because each one is based on a fundamental psychological process that influences behaviour; these influences apply more broadly to general life outside of a sales context. Cialdini identified six of these processes (or what he called principles of persuasion). These principles include Authority, Social Proof, Scarcity, Liking, Commitment, and Reciprocity;[64] two of these are relevant to our discission about Montgomery, Authority and Scarcity.

Obedience to authority

The power of authority to influence actions was demonstrated by an (in)famous series of experiments conducted in the 1960's by an American psychologist, Stanley Milgram.[65] Milgram was interested in the testimony given by Nazi concentration camp guards at the Nuremberg War Crimes Trials, that they were ordinary men who were being obedient to authority and just following orders. To explore this issue, Milgram ran experiments where subjects had to give (what they were led to believe were) electric shocks to another participant. The shocks were nominally administered as part of a learning process, every time the learner (a confederate of the experimenter) made a mistake, the teacher was required to increase the voltage and administer an electric shock. At some point during the process, the learner started to exhibit clear signs of extreme distress and began to ask and then beg for the session to end. The key point of the experiment was the subject's (the teacher) reaction to the situation. All subjects in the study wanted to end the session due to the learner's suffering, but a number were persuaded by Milgram as the experimenter to continue. The machine used to deliver the notional shocks had a dial that displayed the voltage being administered, the top end of which was marked '375 – 420 Danger Severe Shock' and then finally '435 – 450 XXX'. Across the series of experiments, 67 per cent of subjects administered what they believed was a lethal shock. Most subjects were distressed by what they were being asked to do and, in most instances asked, if not pleaded, to be allowed to stop. It should be remembered that Milgram did not hold any legitimate position of power over the subjects; he could not order them to do anything, the subjects were not being paid (other than their expenses) and so were not concerned about losing their

---

[64] Robert Cialdini, *Influence: Science & Practice* (Boston: Allyn & Bacon, 2001), 1-17.
[65] Miles Hewstone et al, *An Introduction to Social Psychology* (Chichester: Wiley, 2015), 264-270.

jobs or losing out on a monetary reward. The conclusion that Milgram drew from the studies was that it was the implicit authority of his status as a professor of psychology (and yes, psychology is a very prestigious subject), and that the experiment was being run as an official university study, that persuaded the subjects to continue. The study has been successfully repeated across different conditions (nationalities and so on) and demonstrates that obedience to authority is a robust psychological process that can exert a strong influence over our actions. Given that Montgomery was operating within a political and military hierarchy, obedience to authority is a useful lens through which his actions can be examined. Let's return to his Tactical Headquarters and look at Montgomery's position.

Montgomery's obedience to authority

Given that he had been pushing so hard for Market Garden, Montgomery, even if he had second thoughts about the wisdom of carrying it out, could not cancel it. This was due to the pressure he was under for the operation to go ahead (implicit authority). The operation had the support of those further up the Allied chain of command, including governmental level.[66] In Britain, both Winston Churchill, (the British Prime Minister) and the British Chiefs of Staff were very keen to see the northern thrust take place. Firstly, there was now an urgent need to overrun the sites for the German V2 weapons now falling on London; after nearly four years of relative security, the British public were experiencing a second Blitz. Montgomery's northern thrust, of which Market Garden was a part, would meet this objective. The British government was also keen to see Britain play a prominent role in postwar European politics. A British-led final push to victory would help to redress some of the humiliations that the nation had suffered earlier in the war (such as the fall of Singapore) and gloss over the fact that Britain was now the junior partner in the alliance. In Washington, Marshall and Arnold were exerting pressure to see FAAA go into action before the war concluded. So, after much discussion, there was now support for the operation within the Allied command in Europe. Eisenhower, who approved the operation on 10th September, was now supportive because it offered the opportunity to gain one of the two crossings across the Rhine that he wished to establish in accordance with his broad front strategy.[67] Montgomery, even if he had

---

66 Clark, *Arnhem*, 113.
67 Powell, *The Devil's Birthday*, 47-48.

wanted to cancel or postpone, would have faced several opponents; besides time was of the essence.

Scarcity principle

Cialdini's Scarcity Principle states that a perceived time pressure or lack of a commodity will generate a demand for the item or make it appear more attractive. The idea is that you are influenced to act so that you do not miss out on the object of interest to you. This principle is used by marketing professionals when they state that a product is a limited edition, or that there are only a few remaining tickets for a concert, or that a sale is only on this weekend. The idea is that the shortage of something or a closing window of opportunity to obtain it should spur you into action.[68] The question is, was Montgomery facing a closing window of opportunity?

Montgomery's urgency

Montgomery knew he needed to act quickly; indeed, it might already be too late. He had conceived the idea for Comet and then Market Garden at a time when it had looked like the German defence was in a state of disarray, something he knew would not last for long.[69] Significant delay in continuing the advance would give the Wehrmacht time to regroup. Montgomery had written to Eisenhower to this effect on 4th September, stating 'time is vital and the decision regarding the selected thrust must be made at once'.[70] The pervading sense of victory euphoria that had generated a wave of optimism at senior levels that the war would be over by Christmas,[71] was now being challenged by the evident signs of the Wehrmacht's reorganisation. Montgomery believed he was facing a closing window of opportunity and needed to act quickly.

Conclusion

Montgomery was facing a difficult situation and under a lot of pressure. The Allied advance was slowing down in the face of stiffening German

---

68 Cialdini, *Influence*, 203-231.
69 Clark, *Arnhem*, 9-10.
70 Montgomery, *The Memoirs of Field Marshal Montgomery*, 271-272.
71 John Colville, *The Fringes of Power – Downing Street Diaries 1939-1955* (London: Wedenfeld & Nicolson, 2004), 483.

resistance just as he was faced with an acute shortage of men and supplies. The argument with his American allies over the limited resources was fierce, with competing strategic proposals being put forward in a hotly contested political climate. It was essentially a zero-sum game. The problem of how to reshape and breathe renewed impetus into the Allied advance, was, therefore, a difficult one. Montgomery also faced other, non-task related pressures; issues caused by the broader situational context. He was subject to serious pressure from the British government to not only clear the V2 sites but also, with one eye on the postwar political situation, to ensure Britain was at the forefront of winning the war while at the same time being frugal with lives and paying heed to the manpower shortage. Added to these pressures was the sense that, with Germany almost beaten, there was a closing window of opportunity to take advantage of the possibilities their weakened but renewing opposition presented.

The outcome of all of this was that, as most commentators have agreed, the operation was put together too quickly.[72] The airborne operations that took place in support of D-Day were of a similar scale, slightly smaller even, and had taken months to plan.[73] The short timescale now for Market Garden imposed a lack of flexibility in the plan and, as we shall see, mistakes were made. Geoffrey Powell, an historian who fought at Arnhem, argues that the lack of time was the key to everything;[74] it certainly had an impact on Montgomery in the way in which he pushed for the operation to go ahead. He would have observed, in the first stage of the OODA Loop, considerable task and situational pressures. The way he responded to these pressures would very much depend on how he made sense of the situation as he moved into the second stage of the OODA Loop – Orientation. This is where his personal filters come into play and in particular his 'difficult' personality, something we will examine in the next chapter.

---

72 Harvey, *Arnhem*, 37.
73 Bennett, *A Magnificent Disaster*, xii.
74 Powell, *The Devil's Birthday*, 243.

2

Montgomery's Orientation – Grip

'I could not have been more astonished'.[1]

Difficult Personality

Montgomery has been described in lots of different ways. He is generally seen as having a 'difficult personality'; Beevor has even postulated he might have been high-functioning Asperger Syndrome.[2] I think this is a suggestion too far; there is, as we shall see, another way of looking at this. Montgomery has been described as the master of the set piece battle; this had certainly characterised his performance in the Western Desert and Italy. By contrast, Market Garden was 'complex, audacious and risky'.[3] It can be argued it was too ambitious, that Montgomery's habitual professionalism had deserted him in this instance. Brooke would later comment that the operation represented a lapse in Montgomery's normally sound military judgement.[4] A key question for us to explore here is what drove this out of character behaviour, this lapse of judgement? This is where we can examine his character. But, before doing so, it will be useful to briefly examine Montgomery's 'difficult personality'.

Montgomery was certainly an unorthodox officer. In terms of the class-ridden, even snobbish attitude that pervaded the British Army in the 1920s, 30s and 40s, and to borrow a rather apt cricket analogy, he was definitely a 'Player and not a Gentleman'.[5] He has been described as 'vain and dictatorial' and 'detested' by his colleagues.[6] Eisenhower would later describe him as a

---

1 Ryan, *A Bridge Too Far*, 94.
2 Anthony Beevor, *Ardennes 1944: Hitler's Last Gamble* (London: Penguin, 2015), 101.
3 Clark, *Arnhem*, 100.
4 Bennett, *A Magnificent Disaster*, 196.
5 Irving, *The War Between the Generals*, 162.
6 Clark, *Arnhem*, 13.

psychopath for whom 'everything had to be perfect' and 'unable to admit to making a mistake'.[7] If he struggled to get on with his British colleagues, he certainly annoyed his American allies.[8]

Nigel Hamilton, his biographer, points to his unhappy childhood as playing a key part in shaping his personality. His mother, Maud, appears to have been unhappy in her marriage and was a 'tyrant', regularly caning the young Montgomery and essentially psychologically 'enslaving' him.[9] Montgomery attests to having had an unhappy childhood, and although he loved his mother, he battled with her throughout his life.[10] Dwelling on this here is unhelpful but the relationship with his family will become relevant later. Hamilton also, for reasons uncertain, spends quite a lot of the opening section of the first part of his biographical trilogy discussing whether Montgomery was homosexual and whether his concern for the men under his command was homosocial. We shall not dwell on this here either, but he did have a deep concern for his soldiers and a concern over morale was a key part of his leadership style. Hamilton goes on to point out that Montgomery's use of the press and public relations was in part driven by the need to be 'seen' by his troops and to keep up morale.[11] It was also, unfortunately one aspect of his behaviour that has laid him open to accusations of vanity and egomania; it was certainly another aspect of his difficult personality.

Hamilton admits that Montgomery was prone to egoism.[12] He charts his growing vanity and immodesty through the course of the Western Desert campaign in 1942-43, highlighting how this increased after the successful battle of Medenine in Tunisia when he beat his erstwhile opponent Erwin Rommel for the second time, and in the process saved Eisenhower's blushes over a faltering American campaign into the bargain.[13] Hamilton does not shy away from discussing Montgomery's eccentricities and pretensions to military grandeur.[14] Ryan points out that by 1944, Montgomery, mainly due to his victory at El-Alamein, was the idol of millions of British public.[15] One source has described him as military deity in that 'He walked in the company

---

7 Ryan, *A Bridge Too Far*, 83.
8 Buckley & Preston-Hough, *Operation Market Garden*, 13.
9 Nigel Hamilton, *Monty: The Battles of Field Marshal Bernard Montgomery* (London: Hodder & Stoughton, 1994), 7.
10 Ibid., 9-11.
11 Ibid., 179.
12 Ibid., 309.
13 Hamilton, *Monty: The Battles of Field Marshal Bernard Montgomery*, 171.
14 Hamilton, *The Full Monty*, 411.
15 Ryan, *A Bridge Too Far*, 83.

of the great captains, his every step a vindication of their living presence' and that he had become 'a rule unto himself and could do no wrong'. Air Marshal Arthur Tedder, Eisenhower's deputy, disliked Montgomery intensely, describing him as 'a little fellow of average ability who has had such a build-up that he thinks of himself as Napoleon'.[16]

Montgomery has also been accused of being a 'glory seeker'.[17] Omar Bradley later stated that the plan for Market Garden came from Montgomery's megalomania and a desire to reap the personal glory that bringing the war to an early close would bring.[18] Brian Urquhart, the Intelligence Officer at British I Airborne Corps agrees. He stated in a 1988 interview that, in his opinion, the plan stemmed from Montgomery's vanity and ambition.[19] This is a little unfair; the successful completion of the operation would have enhanced his reputation but, given the situational pressures we discussed in the last chapter, it would also have allowed the British government to 'punch above its own weight' in the closing stages of the war.[20] Ultimately though, as Lloyd Clark points out, Market Garden, if it came off, would certainly suit Montgomery's ego, profile and desire for influence.

Montgomery's vanity, egomania and 'difficult personality' were clearly an issue and probably underlay his drive to push Market Garden forward. But this is not the whole picture and oversimplifies it. There is another aspect of personality that provides a better explanation of his behaviour and is much more interesting; this is where we need to examine Montgomery's personality more broadly, rather than it just being 'difficult', and one model in particular, the Myers-Briggs Typology.

Personality

The Myers–Briggs Type Indicator (MBTI) is a self-report questionnaire that identifies psychological preferences in how you perceive the world around you, process information, make decisions and structure your life. The information-processing and decision-making aspects make it particularly useful to examine Montgomery's role in devising Market Garden. The psychometric tool was designed by Katharine Cook Briggs and her daughter

---

16 Clark, *Arnhem*, 12-13.
17 Ibid., 333.
18 Buckley & Preston-Hough, *Operation Market Garden*, 205.
19 Baynes, *Urquhart of Arnhem*, 167.
20 Clark, *Arnhem*, 332.

Isabel Briggs Myers based on the psychological theory of Carl Jung.[21] The underlying model has been successfully adapted for use in the defence community in profiling the decision-making and interpersonal styles of key government and military figures.[22] The Myers-Briggs model looks at personality in terms of four dimensions or factors. These factors are quite broad in nature and relate to the following psychological processes: where you focus your attention and draw your energy from; how you prefer to take in information; how you prefer to make decisions; and how you prefer to live your life. These dimensions are bi-polar in nature in that different personality characteristics are associated with either end of the dimension; you can use each side of the different dimensions, but you prefer using one over the other. This preference is your 'default setting' and is where you spend most of your time. This means that the capabilities associated with your preferred style are much more practised and entrenched whilst those that are associated with the other end of the dimension are less developed and thus less effective and efficient. Think about handwriting, you can write with either hand but your writing with your preferred hand is much more fluid, easy and effective. In Myers-Briggs terms, your personality type is made up of your preference for one end of each of the four dimensions. Let's look at each of these four dimensions in turn.

Focus of Attention

Extraversion-Introversion

The first factor in the Myers-Briggs model is where you prefer to focus your attention and draw your energy from. One end of the dimension is called Extraversion and refers to those who prefer to focus on the external world. The other end of the spectrum is called Introversion and refers to those who prefer to focus on the inner world. According to the Myers-Briggs model, if you prefer Extraversion, because you are attuned to the external environment, you are more likely to: communicate by talking; work through problems by doing and discussing; be sociable and expressive; have a breadth of interests; and prefer to have others around you. Conversely, if your preference is for Introversion, because you are drawn to your inner world, you are more

---

21 Isabel Briggs Myers, *Introduction to Type* (Oxford: OPP, 1994), 1.
22 Robert Clark & William Mitchell, *Deception: Counterdeception and Counterintelligence* (London: Sage, 2019), 61.

likely to: prefer to communicate by writing; work through problems by quiet reflection; be private and restrained; have a depth of interest; focus readily; and be interested in facts and ideas. In the model, these preferences are typically referred to by the shorthand of the initial letter of the preference – 'E' for Extraversion or 'I' for Introversion.[23]

Let's turn to Montgomery. To do this, we can look at descriptions of his behaviours at different points in his career, aided by observations recorded by different commentators who observed him in action. The first question to be addressed is where he focused his attention, either on the outer world, Extraversion (E), or the inner world, Introversion (I).

Montgomery's Introversion

We are, unfortunately, starting off with the dimension that we have the least robust evidence for; we do, however, have enough to make a tentative judgement about Montgomery's preference for Extraversion or Introversion. It is tempting to say that his social awkwardness and at times crass behaviour stems from a preference for Introversion. Montgomery, in his younger life, has been described as morose and uncommunicative,[24] shy to the point of muteness with girls.[25] This suggests a preference for Introversion but is insufficient to make a judgement. It is also too simplistic as Introverts, in Myers-Briggs terms, can be quite 'socially skilled' and Extraverts can equally be quite restrained. The model does not conceptualise Introversion and Extraversion in the same way as popular culture does; to underscore this, Extraversion is spelt with an 'a' and not a 'o'. The key to this dimension, in Myers-Briggs terms, is where you focus your attention and draw your energy from, the outer or inner world. It is this aspect that gives us a better line into Montgomery's preference; a couple of good examples provide better evidence for a preference for Introversion. When his wife Betty died in 1937, Montgomery went to the funeral on his own, forbidding any of his family to attend.[26] He also spent many days shut away by himself. He did not want others around him at his time of grief. Later, when he returned to England on leave in 1943 after the Western Desert campaign, he did not want to see any of his family and had planned to spend time on his own. We have seen

---

23 Briggs Myers, *Introduction to Type*, 4.
24 Hamilton, *The Full Monty*, 26.
25 Ibid. *Monty*, 32.
26 Ibid., 259.

that his childhood was not happy, that he had a difficult relationship with his mother, and so there may be an element of this in him not wanting to see his family.[27] It is also suggestive of someone wanting to recharge his batteries in peace and solitude rather than socialising. The evidence, in summary, is not exhaustive, but is suggestive of a preference for Introversion. We will proceed on this basis and double check this later when we look at his overall personality type. We are on much firmer ground with the next dimension, how he took in information.

Taking in Information

Sensing-iNtuition

The next factor the model looks at is how you prefer to take in information or how you tend to perceive the world around you. One end of the dimension is called Sensing (S); this refers to a preference for looking at the here and now and focusing on the practical details of a situation, relying on what your 'senses' tell you. If your preference is for Sensing, you will tend to: value practical applications; be factual and concrete; focus on the present; notice detail; build up a picture methodically; and trust your experience. The other side of the dimension is iNtuition (N), (the model uses 'N' because 'I' has already been used for Introvert); this refers to a preference for looking at the bigger picture and having a future orientation, using your 'intuition'. If your preference is for iNtuition, you will tend to: value imaginative insights; look at the abstract and theoretical; focus on the future; jump around in your thinking and trust inspiration.[28] The next question to consider, then, is how Montgomery preferred to take in information, either Sensing (S) or iNtuition (N).

Montgomery's Sensing

Evidence points strongly to Montgomery having a Sensing preference, indeed, this appears to be his key characteristic or strength. The Sensing preference, as highlighted above, involves a good grasp of detail. Montgomery has been described as the 'master of detail'.[29] He was known for having detailed

---

27 Hamilton, *Monty: Master of the Battlefield*, 275.
28 Briggs Myers, *Introduction to Type*, 4.
29 Hamilton, *The Full Monty*, 635.

knowledge of all the officers under his command; he was called the 'Oracle' during his time commanding Eighth Army in the Western Desert as he knew all the officers from the rank of Major and above.[30] He made a point, upon taking command of Eighth Army, to establish which officers were good and which were not, and to extend this knowledge down to brigade level.[31] Montgomery similarly demonstrated his attention to detail in the First World War. As a Divisional Chief of Staff, he was effective at working through the detail of orders, quick to grasp the need for tactical level reporting using both radio and liaison officers.[32] There are many other examples, but Montgomery seems to fit the detail conscious aspect of Sensing, another aspect is a sense of realism.

A key aspect of Montgomery's behaviour throughout his military career was his 'fundamental streak of realism'.[33] He certainly appears to have looked at situations with a strong sense of pragmatism, which can be seen in his handling of Eighth Army when he took over command. Rather than setting it the sort of ambitious tasks that had got his predecessors into trouble, he set more modest but realistic goals that were within the capabilities of the units he had inherited. Commentators have seen this as caution, but Hamilton argues, correctly in my opinion, that it came from a realistic appraisal of the situation facing him. Montgomery was also pragmatic enough to change his approach at El-Alamein when his original plan was not working. This pragmatism can also be seen in his view of his own intellectual talents, he felt that he did not possess any 'cleverness' but rather 'trained common sense'. Another aspect of his pragmatism was a clarity and precision of thinking.

Montgomery continually demonstrated the ability to boil things down into simple terms. He was noted, during his time in Ireland in 1921, for the clarity and conciseness of his operational instructions. When assuming command in the Western Desert in 1942, Eighth Army had several different plans, Montgomery simplified these down to one: staying and defending the line at Alam Halfa. He was also known for demonstrating an ability to establish clear objectives and setting out how these would be achieved. When teaching at the staff college at Camberley another aspect of his approach was his simplification of tactics and staff organisation,[34] an approach also seen in the training exercises that he organised.

---

30 Ibid., 665.
31 Ibid., 504.
32 Ibid., 119-120.
33 Hamilton, *Monty: Master of the Battlefield*, 196.
34 Hamilton, *The Full Monty*, 166.

Montgomery was recognised as an excellent trainer of men, who was especially adept at creating realistic training rather than hypothetical exercises. He wrote and issued a lot of operational directives and training manuals,[35] which were models of how to translate complexity into clarity, something he certainly had a gift for. He was also keen on capitalising on another form of education, learning lessons from experience.

Montgomery was very keen on conducting 'lessons learned exercises', often sitting down immediately after or even towards the end of a battle or operation to identify key lessons. He then went to great lengths to circulate these thoughts.[36] This meticulousness indicates that Montgomery valued experience, another part of the (S) Sensing preference, and this neatly sums up the difference between a Sensing and iNtuition preference, he very much valued experience over imagination.[37] In summary, it seems that Montgomery had a very clear preference for Sensing when taking in information, the next question we need to address is how he preferred to make decisions based on this information.

Making Decisions

Thinking-Feeling

The third factor looked at by the Myers-Briggs model is how you prefer to understand the information you have taken in and make decisions based on this data. This dimension is linked with the previous Sensing-iNtuition dimension, indeed the two are seen as the 'core' functions; this will become important later, but for now let's focus on this dimension. One side of this dimension is called Thinking (T) and refers to a preference for taking an objective and rational approach to decision-making. If your preference is for Thinking, you will tend to: be analytical and logical; adopt an impersonal approach, be task-focused; be guided by general principles; be reasonable and fair-minded; be seen as 'tough-minded'; and tend not to show appreciation for others. Feeling (F) refers to a preference for taking a subjective and emotive approach. If your preference is for Feeling, you will tend to: be guided by personal values; strive for harmony; assess the impact on people; be compassionate and accepting; and be sympathetic to other

---

35 Ibid., 120.
36 Ibid, ad passim.
37 Hamilton, *The Full Monty*, 166.

peoples' needs.[38] Let's explore Montgomery's preference for Thinking or Feeling. As per the previous dimension, his preference for Thinking is very clear.

Montgomery's Thinking

One of the more frequent terms used to describe Montgomery is 'logical', a key aspect of the Thinking preference. Hamilton, his biographer refers to him as 'supremely logical',[39] making frequent comment about not only his logic but clarity of thought.[40] He goes on to describe Montgomery's time as a Brigade-Major in the First World War as displaying an 'almost psychopathic clarity of mind and decision-making'.[41] His logic and clarity of thought are, along with his grasp of detail, perhaps the other key features of his personality. His logical, impersonal approach manifested itself in other ways, particularly with the way in which he handled emotions.

Montgomery often displayed a lack of emotion in his letters home from the Western Front during the First World War. His discussion of the high level of casualties showed a lack of emotion and his account of visiting the graves of some fellow officers, who he knew well, had a strange dissociative quality. In a similar vein, he would also comment on the destruction of Neuve Chapelle in a dispassionate and calculating manner. Montgomery also frequently talked about 'bottling up' emotions, he was certainly not an outwardly emotional individual. This is evident in his courtship of Betty, his future wife, which was characterised by a no-nonsense approach which lacked emotion.[42] Hamilton speculates that if Montgomery was homosexual, then getting married was a calculated move to further his chances of promotion. We don't want to make too much of this; there is enough evidence, spanning different circumstances, that suggest Montgomery had a detached and impersonal approach to issues. This impersonal approach also found expression in another aspect of a Thinking preference, one that brings us on to another part of his 'difficult personality', a lack of consideration of others.

Another common comment made about Montgomery is his lack of tact and interpersonal understanding.[43] We have already commented on his poor

---

[38] Briggs Myers, *Introduction to Type*, 5.
[39] Hamilton, *The Full Monty*, 443.
[40] Hamilton, *Monty: Master of the Battlefield*, 47.
[41] Hamilton, *The Full Monty*, 73.
[42] Ibid., ad passim.
[43] Hamilton, *The Full Monty*, 197.

relations with his American allies and indeed his British colleagues. As we have seen, he was widely disliked if not detested by different people. His difficulties did not just run to his professional relationships, he was certainly prone to making misjudgements in personal matters. This can be clearly seen, whilst he was in Egypt, in his 'almost paranoid fears for the welfare of David [his son] and his rejection of his family' which 'caused hurt to the members of Monty's family'.[44] An example of this type of behaviour is his order to Mrs Reynolds, with whom David was staying (and not his family) to not allow him to stay with his daughter-in-law, Jocelyn Carver.[45] This was the same woman who had nursed his dying wife in his absence. This lack of consideration for (the feelings of) others can also be expressed in another aspect of a Thinking preference, an intolerance of and criticism of others.

Montgomery tended to make disparaging remarks about other officers, as seen early in his career whilst at staff college in 1920, he was often critical of others.[46] Later, he frequently showed a great deal of immodesty and was often offensive to other officers because he judged them incompetent. The problem is he was invariably right but made these wrong sorts of statements in a setting that should have favoured politeness and social graces.[47]

In summary, it seems that Montgomery had a very clear preference for Thinking when making decisions. The next question we need to address is how he preferred to live his life.

Dealing With the Outside World

Judging-Perceiving

The final factor in the Myers-Briggs model looks at the manner in which you prefer to interact with the outside world. Judging (J) refers to a preference for living a structured, planned and orderly life. If your preference is for Judging, you will tend to: like to have things scheduled; be systematic and methodical in your approach; prefer to plan ahead; like to have things decided; attempt to control situations and events; prefer routine and settled conditions; avoid last-minute stresses; rarely change your mind; and be phased by surprises or changes of plan. Perceiving (P) refers to a preference for living life in an

---

44 Hamilton, *Monty: Master of the Battlefield*, 136.
45 Ibid., 173.
46 Hamilton, *The Full Monty*, 131.
47 Hamilton, *Monty: Master of the Battlefield*, 174-177.

unstructured, open-ended and spontaneous fashion. If your preference is for Perceiving, you will tend to: like spontaneity; be unstructured and flexible in approach; prefer to keep your options open and not 'burn any bridges'; be accommodating to changes of plan and surprises; adapt your approach to circumstances and events; and change your mind easily.[48] We are again on firm ground here as Montgomery appears to have a clear preference for Judging.

Montgomery's Judging

As with the previous two dimensions, common descriptions of Montgomery refer to him in ways that suggest a preference for Judging. Montgomery is typically described as a great organiser, systematic and methodical in his approach, the 'master of the set piece battle', however his critics argue that he was too 'slow and methodical'.[49] He was certainly decisive, prepared to make decisions and generally keen to stick to them once made. Throughout his career he pushed the idea of forward planning, championing the necessity of having a clear plan and sticking to it.[50] Montgomery was also happy to be in control of matters throughout his career. Hamilton suggests the siege warfare in the First World War trenches suited Montgomery, allowing him to exercise his talent for organisation.[51] This ability was readily demonstrated during his time in Ireland with meticulously planned and executed drives against the rebel groups that operated at the time.[52]

A couple of points from his personal rather than professional life, help to reinforce his Judging preference. Montgomery was very keen on good timekeeping; indeed, Hamilton describes him as something of a 'martinet about punctuality'.[53] He also led a very strictly regulated daily routine, with, for example, 30 minutes allotted for breakfast; he was also fastidious about retiring to bed at the appointed time, even on the eve of battle.[54] In summary, it seems that Montgomery had a very clear preference for Judging in terms of how he lived his life.

We have now, a picture of him across all four dimensions, with preferences for Introversion (I), Sensing (S), Thinking (T) and Judging (J).

---

48 Briggs Myers, *Introduction to Type*, 5.
49 Harclerode, *Arnhem*, 16.
50 Hamilton, *The Full Monty*, 243.
51 Ibid., 81.
52 Ibid., 157.
53 Ibid., 136.
54 Ibid., 204.

This allows us to look at all four preferences together and get a more holistic picture of his personality.

Type

Sixteen Combinations

As discussed above, the different dimensions are both independent and seen as dichotomous, in that you have a mutually exclusive preference for one end of the dimension over the other; this means there are sixteen different possible combinations or personality types. Each type is typically referred to in shorthand by the combination of letters that designate the different preferences. For example, if you are an ISTJ, you have the following preferences: Introversion, Sensing, Thinking and Judging. Alternatively, if you have the following set of preferences: Extraversion, iNtuition, Feeling and Perceiving, you are an ENFP.[55] In Myers-Briggs terms the whole is greater than the sum of the parts in that the combinations interact in different ways to produce different personality types; your personality type is not just a summation of your preferences. As we have just seen, there is good evidence to suggest that Montgomery's preferences for Introversion (I), Sensing (S), Thinking (T) and Judging (J) make him an ISTJ.

Within the Myers-Briggs model, if you are an ISTJ, you can be described as 'serious and quiet'. You 'earn success by concentration and thoroughness'. You are 'practical, orderly, matter of fact, logical, realistic and dependable'. You will also 'see to it that everything is well organised'. You 'take responsibility' and 'make up your own mind about what should be accomplished and work steadily, regardless of distractions'. Furthermore, if, as an ISTJ, you 'do not find a place where you can use your gifts and be appreciated for your contributions, you can become rigid about time, schedules and procedures, be critical and judgemental of others, and find it hard to delegate'.[56] Turning to the darker side of things, if you neglect your non-preferred iNtuition and Feeling (remember Montgomery's childhood), you can at times, fail to 'see the wider ramifications of current, expedient decisions.' You can also 'concentrate on logic so much that you don't consider the impact on people' and 'not respond appropriately to other's need for

---

[55] Briggs Myers, *Introduction to Type*, 6-7.
[56] Ibid., 8.

personal rapport and intimacy'.[57] Indeed, under great stress, you may be 'unable to use your customary calm, reasonable judgement and get caught up in imagining a host of negative possibilities';[58] more about this later.

In terms of making contributions to an organisation, if you are an ISTJ, you are likely to: 'get things done steadily and on schedule; be particularly strong on detail and careful in managing it; have things in the right place at the right time; and work well within an organisational structure.'[59] In terms of leadership, you will: 'use experience and knowledge of the facts to make decisions' and 'build on reliable, stable and consistent performance to take charge'. You will also 'respect traditional, hierarchical approaches; reward those who follow the rules while getting the job done' and 'pay attention to immediate and practical organisational needs.'[60] These are the various descriptions of an ISTJ. Let's explore if this fits with Montgomery's overall personality.

Montgomery's Type

The description of an ISTJ, taken as a whole, appears to be a good summary of Montgomery's character, someone described as 'fastidious, obsessively tidy and captivated by detail',[61] to the point of being a 'control freak'.[62] He would be prone to meddling in order to take control of matters, at times issuing orders direct to divisional commanders, bypassing corps commander in the process.[63] Montgomery would describe his approach as 'scientific'. He was determined to 'sort and simplify', and he certainly brought 'clarity, thoroughness and a sense of certainty to operations'.[64] This brought out a level of professional competence that was repeatedly demonstrated throughout his career.

Montgomery, as Brigade Major for 104th Brigade in 1915 'imposed order on chaos', he would end up effectively running the brigade. Later in that war, he would move on to running a division as its Chief of Staff at the age of 30. In May 1940, Montgomery brilliantly handled the night move of

---

57 Ibid., 8.
58 Briggs Myers, *Introduction to Type*, 8.
59 Sandra Hirsh & Jean Kummerow, *Introduction to Type in Organisations* (Oxford: OPP, 1994), 16.
60 Hirsh & Kummerow, *Introduction to Type in Organisations*, 16.
61 Hamilton, *The Full Monty*, 136.
62 Ibid., 567.
63 Hamilton, *Monty: Master of the Battlefield*, 302.
64 Hamilton, *The Full Monty*.

his division in France in the face of the German attack. Within a few days, he would prove adept at managing his part of the evacuation of the British Expeditionary Force (BEF) at Dunkirk. Two years later, he would go on to rejuvenate Eighth Army in four days. Part of this success was due to his intolerance of 'amateurism and indolence'.[65]

Montgomery was ruthless in weeding out officers who he believed were too old, obstinate or incompetent. This was a pattern that he would repeat throughout his career, starting with his early command in India. Conversely, Montgomery supported officers that he thought competent and effective. He thus championed promotion on merit rather than time served or being 'clubable'; he rejected tradition and amateurism in favour of meritocracy and education. His focus was on ensuring all units and formations under his command were competent enough to carry out the tasks assigned to them. The emphasis was on professionalism, his main means of doing so was through 'intelligent rehearsal'; he was very focused on ensuring units engaged in realistic practices before major operations. Montgomery placed a very high value on tactical efficiency and effectiveness. He could not stand inefficiency in others, and was, as we have seen, a model of 'self-contained, focused, cool efficiency'.[66] This is clearly seen in his preparations to defend against the expected German invasion of Britain in the summer of 1940, ordering the removal of wives and families from the invasion area because they would be a distraction to the officers under his command, as well as requisitioning homes and turning them into defensive strongpoints, much to the anger of all concerned. He certainly didn't take no for an answer and rode rough shod over any complaints.

Montgomery could be quite 'super-efficient' in his task focus; indeed, he could be quite autocratic at times. He certainly exuded calm and complete authority and expected orders to be obeyed instantly; his biographer even goes so far as to assert that he had a fixation with discipline and control of others. Another aspect of this side of his personality was that he found it difficult, if not impossible, to admit he was wrong.[67]

On a different note, one of the theoretical aspects of the ISTJ type is perhaps a too strong a focus on the detail and the here and now, with a concomitant difficulty to look at the wider picture. Some commentators suggest this was the case. Bedell-Smith, Eisenhower's Chief of Staff stated

---

55 Hamilton, *The Full Monty*.
56 Ibid., ad passim.
57 Hamilton, *The Full Monty*.

Montgomery 'had never grown above corps or at most, army commander level'. Montgomery's former head of intelligence, Bill Williams, agrees with this assertion, stating that 'Montgomery thought of himself as an army group commander but ... acted more like an army commander. He still involved himself closely in the affairs of Second Army, dictating precisely what was to be accomplished and leaving precious little initiative to his army commander'. Ritchie argues that it was for this reason that Montgomery was not up to commanding a 'highly complex joint operation like Market Garden'.[68]

Montgomery's penchant for meddling makes his lack of involvement in the execution of Market Garden even more puzzling and out of character. So, how can we use our analysis of Montgomery's personality type to explain his out of character behaviour between the end of August and beginning of September 1944? Well, there is an added dimension to the model, a hierarchy to the preferences that will help us to understand this.

Hierarchy

Dominant and Inferior Functions

The Myers-Briggs model asserts that the inner or 'core' functions (Sensing-iNtuition and Thinking-Feeling) exist in a hierarchy of preferences which differs for each type. Understanding how this hierarchy is generated can be fiendishly complicated, so for the sake of brevity and clarity, we will not discuss the process for identifying these here (for a full explanation see Naomi Quenk's *In the Grip*).[69] Simplistically, one of your four preferences for Sensing-iNtuition and Thinking-Feeling forms the 'dominant' function, this is where you spend most of your time; this means it is very well practised and is your key strength. The hierarchical nature of the preferences means that for any personality type, given there is a dominant function, then there is a corresponding 'inferior' function. The inferior function is the opposite of your dominant one and so you spend little time using this approach; this means it is not well-practised and can be seen as your key weakness. The problem arises when stress pushes you into the inferior function, and you remain in its 'Grip'. We have deliberately put this in capitals as it is different from 'gripping' the situation. Grip with a capital 'G', as will be

---

68 Ritchie, *Arnhem*, 257.
69 Naomi Quenk, *In the Grip: Understanding Type, Stress and the Inferior Function* (Oxford: CPP, 2000), 6-9.

explained below, is used in the Myers-Briggs model to describe a very specific psychological state; we will continue to use the capitalised 'Grip' to refer to this state.

Grip Reaction

Of specific interest to our discussion, the Myers-Briggs model suggests that under stress or times of difficulty, you will look to your dominant function (your strength) but as the situation continues, the behaviours associated with this preference become more exaggerated and dysfunctional. If the situation worsens or the stress becomes particularly acute, you then reject the dominant function and swing towards your inferior function. The problem in this situation is that the capabilities associated with that function are not well practised and so the behaviours are dysfunctional. For example, as an ISTJ you would have a dominant function of Sensing, and so you will typically be very grounded, realistic, detail focused, and good at solving problems in the here and now. Under pressure, as the preferred behaviours start to become extreme, you will become more obsessed with detail to the exclusion of the bigger picture. Thus, as an ISTJ you would tend to focus on dealing more and more with solving the immediate and concrete problems. If the problem becomes acute and this approach ceases to work, you abandon the dominant function (that has got you into this trouble) and try something else, switching to the opposite inferior function. Your weak or under-developed iNtuition function means you lose touch with reality, start generating unrealistic plans, lose control over details and become impulsive. Being in the 'Grip' of the inferior function therefore involves behaviours that are atypical of you and, rather unhelpfully, that these atypical behaviours manifest themselves at times of acute pressure.[70] This is where we start to move into an explanation for the key question that we asked about Montgomery's out of character behaviour. Because the Myers-Briggs model examines how your personality type affects how you respond to pressure and stress, it is therefore a good lens through which to view Montgomery's reaction to the situation he faced in August and September 1944. This is especially the case as the model also outlines the events that are likely to trigger a Grip reaction, so we shall re-examine the situation that existed in August/September 1944, to see if these triggers existed for Montgomery.

---

70 Quenk, *In the Grip*, 6-9.

Montgomery's Grip Reaction

Montgomery's Grip Triggers

For the ISTJ, the triggers for the switch to the inferior function and the Grip reaction are: 'others incomplete or sloppy work that affects the quality of their own work, requirement to do things in an inefficient and ineffective way, deadlines, and being asked to change something with no good rationale provided.'[71] Let's examine each in turn.

In terms of a requirement to do things in an efficient and effective way, we have explored in some detail in the previous chapter the debate about which strategy to adopt, a broad front approach, southern thrust or northern thrust. Eisenhower was championing the broad front strategy, Bradley and Patton were pushing for the southern thrust; the northern thrust was, of course Montgomery's preferred solution.[72] In Montgomery's estimation, the northern thrust approach was the soundest military strategy, but he was not alone in this opinion, senior German officers interviewed in 1945 agreed that the northern thrust was the best option for ending the war.[73] By contrast, the broad front approach made less sense from a military perspective; it did, however, make sense from a political perspective as it would not favour one ally over the other.[74] The problem was that Montgomery failed to grasp the political issues.[75] In particular, he could not understand the political imperative of needing to have the larger coalition partner in charge.[76] Eisenhower was clear that American public opinion would not accept holding Patton back in the south and favouring Montgomery in the north.[77] This problem, as shown earlier, was compounded by the fact that 1944 was an election year in the States, making the issue of public opinion loom larger.[78] When Eisenhower explained this to Montgomery, arguing 'public opinion wins wars', the latter replied 'victories win wars … give people victory and they won't care who won it'.[79] Montgomery simply struggled to understand why public opinion

---

[71] Quenk, *In the Grip*, 37-40.
[72] Harclerode, *Arnhem*, 155.
[73] Hamilton, *Monty: The Field Marshal*, 3.
[74] Montgomery, *The Memoirs of Field Marshal Montgomery*, 285.
[75] Powell, *The Devil's Birthday*, 22.
[76] Clark, *Arnhem*, 15.
[77] Hibbert, *The Battle of Arnhem*, 14.
[78] Irving, *The War Between the Generals*, 213.
[79] Ryan, *A Bridge Too Far*, 72.

should drive what he considered to be militarily unsound decisions.[80] He even went as far as writing to the Chief of the Imperial General Staff, Brooke wrote back, however, stating that he could not interfere.[81]

Montgomery was forced to deal with, in his mind, a wholly unsatisfactory situation.[82] He had grave misgivings about the situation, the ongoing debate meant there was no clear plan, and the strategy was unravelling. He was convinced the broad front, or two thrust strategy was wrong, it was simply throwing away the required concentration of effort.[83] In summary, Montgomery's opinion was that the lack of direction or coordination of effort, with two army groups advancing on separate thrusts, was wrong.[84] Thus, the first trigger for Montgomery's Grip reaction, being forced to work in an inefficient and ineffective manner, was present. One of the causal factors of the difficulties he faced was Eisenhower and Montgomery's relationship with him; this brings us on to the next trigger, having to cope with the poor performance of others.

Montgomery was firmly of the opinion Eisenhower lacked control and was failing to sufficiently take charge of the situation.[85] This meant he was not clearly defining what he meant by 'priority', Montgomery even went as far as to confront Eisenhower with evidence of his inconsistent approach at their meeting at Brussels airport on 10th September.[86] Eisenhower, as we have seen, was in a difficult position in terms of needing to consider American public opinion. His main problem, however, was keeping the alliance together. This is why he generally sought compromise solutions that did not satisfy either party.[87] Eisenhower was under pressure from Bradley and Patton to favour the southern thrust and give them priority over logistics, he compromised and gave them some support.[88] This, of course, would impact on Montgomery's efforts, happening at a time when Eisenhower was also assuring Montgomery that he would have priority. As we saw in the previous chapter, this all led to a battle of bids and counter bids for priority over logistics. The problem was exacerbated by the fact that Eisenhower had set up his Headquarters on the French coast at Granville, which was not only 400 miles from the frontline,

---

80 Irving, *The War Between the Generals*, 252.
81 Hibbert, *The Battle of Arnhem*, 25.
82 Montgomery, *The Memoirs of Field Marshal Montgomery*, 266.
83 Ibid., 270-271.
84 Ibid., 286.
85 Ibid., 284.
86 Ryan, *A Bridge Too Far*, 90.
87 Irving, *The War Between the Generals*, 251.
88 Harclerode, *Arnhem*, 23.

but also had poor communications. This meant Eisenhower, who was also immobilised with an injured knee, was essentially out of touch at the very point when speedy decisions were needed. The Market Garden decision was not forthcoming despite Montgomery chivvying Eisenhower three times for it.[89] In summary, there is good evidence to suggest that Montgomery was, in his mind at least, forced to cope with the indecisive and politically motivated 'sloppy' work of others; time was also running out to act.

One of the other triggers of the Grip reaction for ISTJs is deadlines. Although there was not a specific deadline, there was a time imperative. We have already seen that the previously weak German resistance had created a sense of victory euphoria and a general belief that the war would soon be over. The problem was that German resistance was beginning to stiffen, so that a decision needed to be made sooner rather than later.[90] Any real delay would give the Wehrmacht time to regroup. Montgomery was, therefore, aware a key decision point, a deadline in effect, had been reached,[91] and so another trigger was in effect. Realistically, it was probably too late to launch Market Garden by 17th September.[92] Eisenhower needed to take the decision about which strategy to back sooner. Montgomery certainly held this opinion, later stating that the decision came one month too late,[93] the operation should have taken place in August, however, the decision was not Montgomery's to make. He had been up until two weeks previously Land Forces Commander, but had lost this position to Eisenhower, which brings us on to the last trigger, unwanted change.

The final Grip trigger we will look at is when an ISTJ is asked to change something with no good rationale. Montgomery had been, from D-Day onwards, overall Land Forces Commander; this was taken away from him when Eisenhower took over on 1st September.[94] This bitter pill was sweetened to some extent by Montgomery's promotion to Field Marshal, but he was piqued at having to relinquish this position.[95] Ryan goes further suggesting that he was, in fact, humiliated by the change.[96] Ignoring the emotional angle, this move, although it had been planned in advance, made

---

89 Ryan, *A Bridge Too Far*, 80.
90 Montgomery, *The Memoirs of Field Marshal Montgomery*, 272.
91 Hamilton, *Monty: The Field Marshal*, 30.
92 Harclerode, *Arnhem*, 156.
93 Montgomery, *The Memoirs of Field Marshal Montgomery*, 282.
94 Fowell, *The Devil's Birthday*, 22.
95 Buckingham, *Arnhem 1944*, 73.
96 Ryan, *A Bridge Too Far*, 70.

no sense to Montgomery as Eisenhower had to combine his role as Supreme Commander with that of Land Forces Commander. It is hard to disagree with him: Montgomery lobbied hard for one man to act as Land Forces Commander under Eisenhower and that naturally he should take the job; he told Eisenhower this in no uncertain terms.[97] In fact, he felt so strongly about the issue that he even offered to serve under Bradley in that role.[98] In summary, Montgomery simply did not agree with the change and could not see it as a sensible arrangement. Thus, nearly all the ISTJ Grip triggers were in effect. The final question to be answered is did this push Montgomery into the Grip of the inferior function and if so, what behaviours did he exhibit because of this?

Montgomery's Grip Behaviours

To determine whether Montgomery was in the grip of his inferior function we need to examine the extent to which he was finding the situation stressful, whether he exhibited any exaggerated dominant Sensing behaviours and then any of the ISTJ Grip behaviours (losing touch with reality, generating unrealistic plans, losing control over details and becoming impulsive).

Firstly, most commentators agree that Montgomery was 'furious' about the lack of direction and coordination.[99] He was becoming increasingly frustrated and perplexed about the logistics crisis and Eisenhower's lack of contact with him.[100] Added to his frustration was the difficulty of trying to communicate with Eisenhower given the poor communications set-up at Granville.[101] Montgomery's frustration, and indeed despair, are evident in the letters he wrote at the time to people like Brooke. For example, one letter to the CIGS was a long discourse complaining about the situation, stating that Eisenhower 'keeps saying that he has ordered that the northern thrust to the Ruhr is to have priority, but he has NOT ordered this'. He would go on to complain with feeling about the supply situation, adding two exclamation marks in the margin of the typewritten letter.[102] He was also becoming increasingly tetchy; one example illustrates this nicely. General Henry Crerar, the commander of the Canadian First Army did not attend a conference

---

97 Irving, *The War Between the Generals*, 251.
98 Ibid., 304.
99 Buckley & Preston-Hough, *Operation Market Garden*, 206.
100 Hamilton, *Monty: The Field Marshal*, 36-37.
101 Clark, *Arnhem*, 16.
102 Hamilton, *Monty: The Field Marshal*, 43.

called by Montgomery because he chose to attend a memorial service for the Canadian soldiers killed in the Dieppe raid in 1942. Montgomery was furious and snapped, treating Crerar to a vitriolic dressing down when he did finally arrive at his Headquarters.[103] He would later write a letter of apology, but the frustration and bad temper can be seen as signs of someone suffering from stress. His frustration would find other targets; he made disparaging remarks about the Dutch and Belgian units under his command, referring to them with 'such unbridled sarcasm [that] betrayed the very frustration Monty felt'.[104] Montgomery found the politicking exhausting. He had been worn down by the infighting during the invasion of Sicily,[105] and the same response arose again; he wrote to Brooke on 12th September stating: 'I feel somewhat exhausted by it all'.[106] There is good evidence to suggest, then, that by early September, Montgomery was experiencing stress. We next need to turn to the exaggerated behaviours associated with overworked Introverted Sensing.

It was about this time that Montgomery increasingly isolated himself at his Tactical Headquarters and effectively removed himself from the Allied chain of command, all contact was maintained via intermediaries.[107] This is the same behaviour that he displayed when his wife Betty died and suggests a retreat into his inner world. Montgomery also became more focused on trying to tackle the problem. He certainly bemoaned the lack of what he called 'grip' (not to be confused with the Myers-Briggs term 'Grip') being shown by others and believed that without it the Allies were losing the chance to win the war quickly.[108] As we saw in the previous chapter, he sent multiple messages to Eisenhower with more and more urgent appeals to come to a decision, making repeated and ill-judged attempts to persuade him of the need to prioritise the northern thrust.[109] This all suggests someone who was working hard to resolve the situation in (a somewhat overwrought) Sensing mode. One manifestation of this can be a fixation on detail or a specific task. It seems rather odd that at the time that this was happening Montgomery became obsessed with a second portrait that the artist James Gunn had painted of him. Hamilton argues that 'Gunn's portrait now became an almost obsessive bee in his bonnet. Letter after letter now referred to the portrait – "the best

---

103 Hamilton, *Monty: The Field Marshal*, 34.
104 Ibid., 13.
105 Hamilton, *Monty: Master of the Battlefield*, 272.
106 Hamilton, *Monty: The Field Marshal*, 58.
107 Ritchie, *Arnhem*, 101.
108 Hamilton, *Monty: The Field Marshal*, 43-45.
109 Hamilton, *Monty: The Field Marshal*, 34.

picture he [Gunn] has ever painted; it is definitely superb"'.[110] In summary, this suggests that frustrated and stressed by the situation, Montgomery doubled down on his dominant Introverted Sensing function but could not solve the problem. So, did he then fall into the 'Grip' of his inferior function?

It seems that Eisenhower's decision, in late August, to split his force into two separate thrusts, one in the south under Bradley and one in the north under Montgomery pushed the latter nearly to the edge; it was 'nothing less than disastrous' as far as he was concerned.[111] Brooke agreed, noting in his diary that the 'Eisenhower's plan is likely to add another three to six months to the war'.[112] We would argue that the actual edge, the tipping point came a few days later. We have described how Montgomery sent a message to Eisenhower on 4th September again urging one 'full blooded thrust' and asking his boss to come and see him.[113] The tone of the message was arrogant and insolent, practically dictating what Eisenhower should do.[114] Eisenhower was understandably furious,[115] and replied the next day but the communications were so poor that Montgomery received the second half of the message on the 7th and the first half on 9th September.[116] The message basically reiterated Eisenhower's stance, and put forward not one but five objectives, the seizure of the Ruhr and the Saar, along with the ports of Brest, Le Havre and Antwerp.[117] Montgomery vehemently disagreed with the signal,[118] and most likely this was the tipping point that pushed Montgomery into the Grip of his inferior function. Hamilton agrees, concluding that the two messages 'tipped the scales of Monty's otherwise profoundly professional and generally deeply realistic mind'.[119] If this is the case, then he would display the behaviours associated with an ISTJ in the Grip; loss of touch with reality, generation of impractical ideas, impulsiveness and loss of control over details.

In terms of impulsiveness, Montgomery had been slowly moving from cancelling Comet and directing the advance away from Arnhem to suddenly

---

110 Ibid., 33.
111 Ibid., 7.
112 Ibid., 5.
113 Montgomery, *The Memoirs of Field Marshal Montgomery*, 271.
114 Ryan, *A Bridge Too Far*, 79.
115 Baynes, *Urquhart of Arnhem*, 82.
116 Montgomery, *The Memoirs of Field Marshal Montgomery*, 272.
117 Hamilton, *Monty: The Field Marshal*, 37.
118 Ryan, *A Bridge Too Far*, 84.
119 Hamilton, *Monty: The Field Marshal*, 41.

scaling up the operation substantially to Market Garden.[120] Montgomery 'astonished his own operational and planning staff' when he announced this on 9th September. 'Far from closing down the British Arnhem operation Montgomery suddenly insisted it be remounted' and not only remounted but with the 'entire available strength of First Allied Airborne Army, including British, American and Polish troops'.[121] The sudden change of plan took people by surprise.

The proposed operation was imaginative. The senior officers in FAAA were taken aback when briefed on it. They were astonished by its boldness and scale.[122] Bradley would comment it was 'one of the most imaginative [plans] of the war'.[123] We have already seen his comment on Montgomery wobbling into SHAEF with a hangover. The problem was this operation broke with all of Montgomery's principles of war: logistical back-up; adequate reserves; application of firepower; and thorough preparation.[124] In devising the plan, Montgomery demonstrated a strange lack of realism.[125] As already discussed, it was just too ambitious and, to our mind, indicative of an ISTJ thinking in a poorly developed iNtuition mode. Such a mindset also involves a lack of concern with practical matters and lack of engagement with the real world.

One of the other facets of Montgomery's behaviour in relation to Market Garden, that often attracts comment, is his uncharacteristic lack of involvement in the execution of the operation. Montgomery and his staff had nothing or very little to do with the detailed planning and conduct of the operation; Montgomery would make only one intervention, unsuccessfully, to get Brereton's decision of one drop on the first day reversed.[126] Powell has commented 'it is strange that Montgomery, having staked so much on Market Garden, remained so uncharacteristically remote from its actual execution, once the operation had been launched'. He proceeds to quote Montgomery's intelligence chief who observed 'a lack of grip, surprising in Montgomery' and that he appeared to let things go their own way.[127] This corresponds with Harclerode's comments suggesting his lack of involvement

---

120 Ibid., 43.
121 Ibid., 40.
122 Harclerode, *Arnhem*, 47.
123 Hamilton, *Monty: The Field Marshal*, 56.
124 Hamilton, *Monty: The Field Marshal*, 56.
125 Ibid., 73.
126 Bennett, *A Magnificent Disaster*, 21.
127 Powell, *The Devil's Birthday*, 239.

was 'unusual because he was well known for taking too close an interest on occasion when, as an army or army group commander, he should have left matters to his divisional or corps commanders'.[128] Montgomery did not consult with his staff or issue any specific instructions to the two armies under his command when he gave the order for Market Garden. Bennett agrees with his fellow commentators that this was 'uncharacteristic of the man, as was the dearth of interventions over the following week'.[129] A good example of this is Montgomery's failure to go to Nijmegen to 'ginger up' Horrocks when XXX Corps was moving too slowly.[130] This was something he had done repeatedly before, choosing instead to stay in 'near seclusion' at his Tactical Headquarters.[131] Montgomery would later himself admit that he should have stepped in during the planning of Market Garden.[132]

The consensus seems to be that Montgomery did not meddle or even involve himself in the planning and conduct of Market Garden, and that this was highly unusual for him. We would conclude that this was one of the behaviours that came from his under-developed iNtuition and is characteristic of an ISTJ in the Grip of the inferior function. It was this Grip reaction that prevented him from taking a realistic view of the situation and cancelling the operation. Instead, he went for the gamble that was Market Garden. A gamble, that was driven by his Grip reaction.

Conclusion

This chapter set out to answer one of the key questions posed at the beginning of this book. Why was it that Montgomery acted out of character regarding Market Garden? As we have seen, the operational concept was bold, imaginative and out of character for him. It was a marked departure from his usual methodical approach. His lack of involvement in the operation itself was also unusual. We have argued that this uncharacteristic behaviour can be explained by the Myers-Briggs model.

Montgomery's personality type appears to be ISTJ. This fits with the descriptions of him as a great organiser, methodical, detailed, and cautious; the master of the set piece battle. This was all driven by his dominant Sensing function. By early September 1944, he was frustrated by the loss

---

128 Harclerode, *Arnhem*, 173.
129 Bennett, *A Magnificent Disaster*, 9.
130 Hamilton, *Monty: The Field Marshal*, 87.
131 Ryan, *A Bridge Too Far*, 85.
132 Harclerode, *Arnhem*, 174.

of his influential role as Land Forces Commander and was now a man 'in a hurry'.¹³³ Unfortunately, as time and opportunity began to slip away, and he was forced to deal with what he saw as the inefficiency and incompetence of others, he was triggered by this into his inferior iNtuition function. This is where in reaction he ignored the realities of the situation and came up with the gamble that was Market Garden. His uncharacteristic behaviour was driven by his personality and particularly his personality type's specific form of the Grip reaction. Being in the Grip, he lost his customary grip.

In conclusion, Montgomery's reaction to the pressures he faced pushed him into the Grip of his inferior function, leading him to abandon his usual analytical logical approach. The psychological conflict he was experiencing thus meant he was not thinking clearly and methodically. We will examine the impact of this on Montgomery at the Decide stage of the OODA Loop in more detail in the next chapter.

---

133 Bennett, *A Magnificent Disaster*, 9.

3

Montgomery's Decision – Avoidance

'I don't think your resistance people can be of much use to us.'[1]

Coping

Montgomery's Tactical Headquarters, Belgium, 16th September 1944

A good example of Montgomery's coping strategy is his meeting with Bedell-Smith, Eisenhower's Chief of Staff at SHAEF on 16th September. The American general had flown over, along with Kenneth Strong Eisenhower's Chief of Intelligence, for urgent discussions. The latest Ultra decrypt had indicated remnants of 9th and 10th SS Panzer Divisions were regrouping in the Arnhem area and thus posed a threat to Market Garden.[2] Bedell-Smith suggested dropping an extra airborne division to help counter the threat, but Montgomery dismissed the suggestion and the report. Referring to a Second Army intelligence summary, he replied in his view, both SS units were probably not battleworthy.[3] With only two days to go before the launch of Market Garden, it was too late to make any changes to the plan; they would have to make the best of what they had.

Decisional Conflict

This meeting opens a fascinating window into Montgomery's mindset just a few days before Market Garden. Dismissive of challenge, he was obstinately sticking to the plan. We have explored the external pressures on Montgomery

---

[1] Ryan, *A Bridge Too Far*, 86-87.
[2] TNA, Ultra DEFE 3/221
[3] Bennett, *A Magnificent Disaster*, 59.

and how these pressures pushed him into his Grip reaction, where he was experiencing a good deal of frustration and psychological conflict. In this chapter we will explore how he dealt with this conflict, using a framework that examines the different coping strategies that people use when dealing with difficult situations – the Decisional Conflict Model.

The Decisional Conflict Model (DCM) was originally developed as a way to describe and explain people's different responses to warnings of natural disasters such as hurricanes and tornados; why do some people ignore the warning whilst some take heed and act early, and others panic at the last minute?[4] The model suggests that if you are faced with a difficult decision, you will adopt one of five basic coping patterns for dealing with the situation. Each coping pattern is associated with differing levels of information-processing and problem-solving activity thus affecting your ability to effectively handle the situation. You are more likely to make a poor decision if you adopt one of the first four coping patterns as they involve a premature closure of your information-processing and problem-solving activity. Conversely, you are more likely to make a better decision if you adopt the fifth coping pattern, called Vigilance, as, by contrast, it is associated with more extensive problem-solving activity, (you have been 'vigilant' in your information-processing).[5]

The DCM has been applied to a range of other issues, such as reactions to health campaigns,[6] and more importantly for us, how individuals cope with the psychological stress of decision making amidst international crisis situations and military actions.[7] For example, one research study examined whether the outcome of US Presidential decision-making during 19 international crises post Second World War were related to Vigilant information processing. Bibliographic sources describing the decision-making process in each crisis were content analysed and rated for Vigilant information-processing. Experts rated the crisis outcomes in terms of their effect on US vital interests and on international conflict. Results indicated that crisis outcomes had more adverse effects on US interests and international conflict when the decision-making process met less of the criteria for Vigilant

---

[4] Irving Janis & Leon Mann, *Decision Making: A Psychological Analysis of Conflict, Choice & Commitment*, (New York: Free Press, 1979), ad passim.
[5] Irving Janis & Leon Mann, 'Coping with Decisional Conflict', *American Scientist*, 64, (1976), 657-667.
[6] Janis & Mann, *Decision Making*, ad passim.
[7] Philip Tetlock, 'Identifying victims of groupthink from public statements of decision makers', *Journal of Personality & Social Psychology*, 37, 8, (1979), 1314-1324.

information-processing.[8] This evidence, although not exhaustive provides support for the model, which we can examine in more detail.

Situational Assessment

The specific coping pattern that you will adopt depends on your appraisal of the situation. More specifically, your perception of three factors: your awareness that serious risks are involved in the situation; your belief that a workable solution to the problem is possible; and your assessment that there is adequate time to work up and implement the required solution [9]

If you believe that there are clear risks in the situation, but, there is a possibility that you can find a viable solution and there is sufficient time available for you to work up a plan and act on it, you will be in the Vigilant coping pattern and so engage in purposeful, robust information-processing and decision-making. By contrast, if you decide that there aren't any risks, or a workable solution can't be found, or you don't have enough time, you adopt one of the other four coping patterns and thus engage in less effective information-processing and decision-making. The starting point, then, is your perception of risk.

Risk

You are likely to experience psychological (or decisional) conflict if you acknowledge that there are risks, and therefore potential losses, in the situation you face. If you assess that there are no serious risks associated with continuing with your planned course of action or sticking with the status quo, the Decisional Conflict Model predicts you will adopt a coping pattern that it calls Unconflicted Adherence.[10]

Unconflicted Adherence

Your assessment that no significant or meaningful risk is present means that you do not experience psychological or decisional conflict – this is the 'unconflicted' part. Adherence means that you stick with Plan A or the

---

8 Gregory Herek et al, 'Decision making during international crises. Is quality of process related to outcome?', *Journal of Conflict Resolution*, 31, 2, (1987), 203-226.
9 Irving Janis & Leon Mann, 'Emergency Decision Making: A Theoretical Analysis of Responses to Disaster Warnings', *Journal of Human Stress*, 3(2), (1977), 35-48.
10 Janis & Mann, 'Emergency Decision Making', 35-48.

status quo. In this instance, you are likely to engage in little, or indeed, no meaningful search for and processing of information, because you feel that there is insufficient risk to warrant the effort. Let's put this back into the model's original context, reactions to warnings of natural disasters, to explore how the process works.

Imagine you are on holiday in southern Florida in the United States. You are in the bar of your hotel when you hear on the TV news that a major hurricane will hit the eastern seaboard of the United States in three days. The government is warning people to act. According to the Decisional Conflict Model, the first question you ask yourself is: am I at risk? Looking at the news item in more detail it becomes apparent that the hurricane will hit the coast hundreds of miles to the north, in North Carolina. You are not at risk, and you can happily stay where you are. There is no decisional conflict, and you can enjoy your Unconflicted Adherence whilst sipping your Pina Colada. In this instance, your assessment is correct and the coping pattern you adopt is appropriate for the situation. The circumstances, or your assessment of them could, of course, be different and you would then adopt a different coping pattern. If you assess that there are indeed risks associated with your present course of action, but there are no risks linked with the most obvious, readily available alternative, or another standard approach, you will adopt the second coping pattern, called Unconflicted Change.[11]

Unconflicted Change

You still do not experience decisional conflict in the Unconflicted Change pattern; you remain unconflicted, because you can easily change to another course of action. The key point is that, in both cases, you are unlikely to engage in any meaningful problem-solving activity. Your perception of risk might be correct or incorrect, either way, your decision-making will be driven by this assessment. Let's return to the hotel bar. This time the circumstances are different. The hurricane is going to hit southern Florida, you are in clear danger. There is risk. You need to act, is there a quick win? You had been musing about cutting your trip short and indeed you have already been looking up flights for tomorrow. It looks like there is a seat on a flight leaving in three hours at no extra cost. One more click on your phone, and you are

---

11 Janis & Mann, *'Emergency Decision Making'*, 35-48.

on the flight. You have changed the plan but the ease of finding the solution means that you don't really experience any decisional conflict.

The next question we need to explore, using the DCM as a framework, is Montgomery's perception of the strategic situation at the end of August/beginning of September and the risk that he would not be able to attain his objective of not only bringing the war to a close quickly but also ensuring Britain remained a prominent actor in the war effort by getting his northern thrust approved. We saw in the last chapter Montgomery was faced with a difficult strategic situation and was experiencing psychological conflict. His Grip reaction meant that he was not thinking as robustly as he typically did, so how did this affect his risk assessment?

Montgomery's risk calculus

The first point we need to consider here is the general sense of victory euphoria permeating all levels of SHAEF, neatly encapsulated in the 'end of the war is in sight' SHAEF Intelligence Summary of 26th August.[12] This would have had a positive effect on Montgomery's risk appreciation, but it also had a negative impact on the way the available intelligence was interpreted.[13] So, let's try and put ourselves into his shoes again and revisit his Tactical Headquarters and look at the problem as he saw it as he works through his situational assessment.

Montgomery's Tactical Headquarters, 15th September 1944

Montgomery had just been given the latest update by Bill Williams, his Chief of Intelligence. Montgomery had a range of intelligence sources available to him, especially about the German presence in the operational area. One source was Ultra, the top-secret code breaking effort at Bletchley Park that allowed the Allies to read Wehrmacht communications in almost real time. Ultra was a solid and trusted source. Ultra decrypts, received on 4th and 5th September, had indicated that remains of a weakened German Panzer Corps was moving into the Arnhem-Nijmegen region,[14] indeed, Bletchley Park had been tracking the movement of II SS Panzer Corps from Normandy

---

12 Clark, *Arnhem*, 9.
13 Powell, *The Devil's Birthday*, 41.
14 DEFE 3/221, XL 9245, 6 September 1944.

over the last few weeks.¹⁵ More worryingly, by 7th September it was clear that OKW, the German High Command expected an Allied advance in the Arnhem area.¹⁶ Between 9th and 14th September one of Field Marshal Walter Model's (the senior German officer in the west) intelligence officers issued warnings about a Second Army breakout towards Nijmegen, Arnhem, Wesel and the Ruhr, including use of an airborne operation. He even went so far as to conduct a paper exercise where he wrote an order from Eisenhower to Dempsey that essentially set out the strategic aim for Market Garden.¹⁷ The fact the German High Command was aware of Allied intentions to commit to a northern thrust through Arnhem was confirmed in another Ultra decrypt on 15th September.

Ultra was not the only intelligence source pointing to a stronger than expected German presence threatening the intended line of advance, Dutch resistance had also identified the presence of SS Panzer units around Arnhem.¹⁸ Williams, Dempsey and Montgomery had all read these reports. This information had been passed on to 1st Airborne Division, which reported in its intelligence summary that 'one of the broken Panzer divisions has been sent back to the area north of Arnhem to rest and refit'.¹⁹ Crown Prince Bernhardt of the Netherlands had been so alarmed by Dutch underground reports about German armour in the operational area that he had flown to see Montgomery on 7th September to raise his concerns; it had been a disappointing meeting for him.

The problem was that the Dutch resistance, and hence their reports, were not trusted by the British since their networks had been penetrated and compromised by the German counter-intelligence service, the Abwehr, between 1941-43 in what had become known as the Englander Spiel. In this operation, Special Operations Executive (SOE) agents parachuting into Holland were captured by the Germans, these agents, (dutifully omitting their security codes to warn they were under duress) were made to send radio messages back to England to encourage further drops. These warnings were ignored and a total of 54 agents sent into Holland were promptly captured. The integrity and credibility of the Dutch underground was undermined

---

15 Powell, *The Devil's Birthday*, 42.
16 Ritchie, *Arnhem*, 130.
17 Clark, *Arnhem*, 62.
18 Powell, *The Devil's Birthday*, 42.
19 TNA WO 171/393, 1st Airborne Division Planning Intelligence Summary No.2, 1st Airborne Division War Diary, September-December 1944.

when the Allies discovered what had happened.[20] This is one of the reasons why their latest intelligence reports were, despite their relevance and immediacy, dismissed. Montgomery shared this scepticism. At the end of the meeting with Bernhardt he dismissed the reports, commenting: 'I don't think your resistance people can be of much use to us. Therefore, I believe all this is quite unnecessary'.[21] Despite his dismissal of these reports, Montgomery was probably beginning to sense the Wehrmacht was steadily regrouping in the operational area.

Within a short space of time of Second Army resuming the offensive on 6th September, it had become clear that the German defences facing it had been reinforced and strengthened. It was the slower than hoped for progress of XXX Corps, caused by stiff German resistance, that had led him to cancel Comet as it was feared it would not be able to link up with 1st Airborne in time.[22] He was aware, then, that German forces in the area possessed a renewed fighting capability and therefore posed a risk to the operation; this was troubling Bill Williams as well.

Williams was a worried man. He had read various intelligence estimates with growing concern, one dated 12th September, clearly stated that in the Arnhem-Nijmegen area, elements of: '1st, 9th, 10th and 12th SS Panzer Divisions have all been seen in small pockets'.[23] These reports had followed on from one received two days earlier that indicated the presence of 9th SS Panzer Division in the Arnhem area.[24] Williams had been so concerned at the time that he briefed Montgomery to this effect. Now, however, the picture was more unclear. Other estimates seemed to temper the more alarming reports: 'there is no reason to suppose any of these formations is operating as or in a condition to operate as a division at the present time'.[25] The 9th and 10th SS Panzer Divisions were part of the II SS Panzer Corps, that Williams knew had been roughly handled in the Normandy campaign and were understrength; they had clearly been sent to the Arnhem area to recuperate. Despite this, these German units were still likely to pack a powerful punch and had been specially trained for anti-airborne operations.[26] Williams was,

---

20 Powell, *The Devil's Birthday*, 102.
21 Ryan, *A Bridge Too Far*, 86-87.
22 TNA, WO 285/9, Dempsey Papers, Personal War Diary as Commander 2nd Army, 'First 100 Days'.
23 LHCMA, Dempsey Papers, British 2nd Army Intelligence Summary, No. 101, 13 September 1944.
24 TNA, WO 208/3575, Brigadier E.T. Williams, 'Notes on the use [of Ultra],' 5 October 1945, 2.
25 Ibid., 2.
26 Kershaw, *It Never Snows in September*, 38-44.

therefore, becoming increasingly concerned about the strength of resistance that Market Garden and 1st Airborne might face. He was not alone. A few miles away, Miles Dempsey, had been receiving similar reports and was now concerned about the German presence in the area; he had gone as far as to write in his diary: '[The] enemy ... appreciates the importance of the area Arnhem-Nijmegen ... Are we right to direct Second Army to Arnhem?'[27] He had been going to propose Wesel as the axis of advance at his meeting with Montgomery on 10th September but was forestalled by the latter showing him the War Office signal about recent V2 attacks on London. The British did not have a monopoly of concern about the growing strength of German resistance; even further away, alarm bells had also been ringing at SHAEF, prompting Bedell-Smith's meeting with Montgomery.[28]

Risk Appreciation

Looking critically at the situation, it seems clear that both Dempsey and Montgomery were aware of the Wehrmacht's reorganisation in the operational area and that this represented a threat to the success of the northern thrust, indeed, this renewed threat was the reason why Montgomery had cancelled Comet.[29] Roger Cirillo argues that senior Allied officers, including not only Montgomery but also Eisenhower, Brereton, Leigh Mallory and Tedder were sufficiently aware of the risks to Market Garden to stop it but didn't, 'the possibility of failure seemed less than the consequences of cancellation'.[30] A senior Dutch officer has pointed out the British seemed to be reversing Napoleon's famous maxim about fighting only when there was a 75 per cent chance of success and were instead going forward with only 25 per cent chance of success, leaving 75 per cent to fate.[31] Montgomery was, therefore, taking a risk, if not a gamble.

In summary, we would suggest that Montgomery was aware of the risk that his northern thrust would not succeed and so answered the first question in the DCM process in the affirmative. Having accepted the risk, according to the model, he would have moved on to the next question, was there a viable solution to the problem? We, therefore, need to consider feasibility.

---

27 TNA, WO 285/9, Dempsey Papers, Personal War Diary as Commander 2nd Army.
28 Harclerode, *Arnhem: A Tragedy of Errors*, 39.
29 Ritchie, *Arnhem*, 131.
30 Cirillo, *Market Garden and the Strategy of the Northwest Europe Campaign*, 57.
31 Ryan, *A Bridge Too Far*, 150.

Feasibility

Let's return to your Florida hotel and the approaching hurricane. If you assess that there are risks associated with the different available courses of action, you will experience decisional conflict – the agony of choice. Let's change the scenario again and say that the hurricane is going to hit the coast right where you are in the next 12 hours. The problem is that your partner is ill in bed and is not fit to travel easily. There is now risk. In these circumstances you will then be prompted to search for different solutions and move on to the next situational assessment or question in the DCM model; whether there is a viable alternative that can be found. If your answer to this assessment is negative (there is no way that your partner can travel), you will move into adopting the third coping pattern, Defensive Avoidance. In this condition, you face dilemmas – this is the decisional conflict. Let's say that there are alternatives, but they are not great. The hotel has a shelter that has been designed to be storm proof, but it is unlikely to be strong enough to survive the hurricane without damage and threat to life. Staying put is an option but not an attractive one. Frantic ringing round reveals that you might be able to charter a private ambulance to collect you and your partner and drive you out of harm's way. The problem is that, given the demand, and the lateness of the hour, the ambulance company can't guarantee to get to you in time. Even if they do, you might get caught in the open by the hurricane. Again – problematic option with unknowns. In this situation you accept that the situation involves risk and costs but believe there is no real, viable solution. You adopt the coping pattern of Defensive Avoidance.

Defensive Avoidance

Your adoption of the Defensive Avoidance strategy can take one of three different forms. The first is Bolstering. This is where you settle on the least objectionable alternative and boost its attractiveness by wishful thinking, making the best of a bad job. The hotel is storm proof, and if you can get some help, you should be able to move your partner downstairs. You might be alright. Alternatively, you can engage in Scapegoating, this is where you let the decision be taken out of your hands by passing responsibility to (and then if necessary, blaming) someone else. You ask the private ambulance company for an assessment of the chances of them getting to you. The calm voice on the other end of the phone states that he is confident; that's good enough, you decide to put your faith in them. The third form of Defensive Avoidance open to you is Procrastination; this is where you dither so much that you fail to decide. In our hurricane disaster scenario, you are so frozen by indecision that you neither move to the shelter nor go for the ambulance option.

The exact form of Defensive Avoidance that you adopt depends on previous experience and the conditions specific to the situation; for example, you might adopt the ambulance option if you have successfully used one before, alternatively, it will not be a viable option if the company does not have any ambulances available.[32]

A key part of this DCM coping strategy is the need to cope with the psychological strain that your perception of the situation has caused; you know there are risks but you don't think that there is a feasible solution. This is the Decisional Conflict in the model, the stress that needs to be coped with. According to a model developed by Lazarus and Folkman, these coping efforts are aimed at dealing with two issues, the problem itself and the emotions aroused by the situation. Problem-focused coping is defined as those activities that are aimed at resolving the issue that is causing the psychological strain. Emotion-focused coping is aimed at managing the stress or psychological strain that you experience. In terms of managing the situation, problems may then occur because your attempts to manage your emotions will compete for your attention with the need to focus on solving the problem. The problem is your information-processing capacity is limited; attentional capacity is a made up of a finite extent of information-processing resources. Emotion-focused coping activities will therefore take up part of this limited capacity and so leave less attentional capacity for problem-focussed coping.[33] In DCM terms, the dysfunctional coping strategy of Defensive Avoidance involves more emotion-focused behaviours and less problem-focused activity and so will impact on the quality of your decision making.

The next question we need to address, then, is Montgomery's assessment of the viability of finding an alternative solution to his northern thrust. Montgomery's judgement on the feasibility of his northern thrust would have rested on several factors, the first of which was the supply problem and Montgomery was, if nothing else, a logistics expert. We have already discussed the logistics problem at length in the first chapter, but to recap,

---

[32] Janis & Mann, *'Emergency Decision Making'*, 35-48.
[33] Susan Folkman, 'Dynamics of a stressful encounter: Cognitive appraisal, coping, and encounter outcomes', *Journal of Personality & Social Psychology*, 50, 5, (1986), 992-1003.

let's return to his Tactical HQ to look at the situation from Montgomery's perspective as he considers his options.

Montgomery's problems

We can only imagine Montgomery's continuing despair as he reviewed his logistics situation. The leading elements of Second Army, the Guards Armoured and 11th Armoured Divisions, were halted at Antwerp and Brussels respectively.[34] Eisenhower's decision to push forward with two thrusts meant he would not have the priority over supplies needed to undertake Market Garden and exploit beyond the Rhine. The viability of the northern thrust was, therefore, in doubt because of the lack of supplies but he was hopeful a successful completion of the operation would force Eisenhower to change tack and support him with more supplies.[35] Logistics were not the only thing Montgomery was short of; there was also a manpower crisis.

Montgomery's infantry battalions, as we have seen, were chronically under-strength. This lack of manpower meant that there would be limited resources with which to exploit the bridgehead once the bridge at Arnhem had been seized. His limited manpower threw another big question mark over the viability of the northern thrust especially in view of the apparently increasing ability of the Wehrmacht to rush reinforcements into the region. The proximity to the German border and the efficacy of the German railway network meant that the Wehrmacht could likely move reinforcements into the battle quite easily.[36] The problem would be compounded by any delay in XXX Corps' advance, that would give the Germans more time to push reinforcements into the area.[37] The British advance would need to proceed along a single highway, in places a raised road with little room for manoeuvre, a corridor which was also narrow and easily cut. The key issue, then, about German strength in the Arnhem-Nijmegen area was not just the extant strength of the German forces there, but the assessment of whether those forces could be reinforced fast enough to stop or slow down XXX Corps.[38] This was not the only problem associated with XXX Corps' rate of advance.

---

34 Clark, *Arnhem*, 10.
35 Mead, *General 'Boy'*, 117.
36 TNA, AIR 37/1217, Operation Market, 1st Airborne Division Planning Intelligence Summary, No.2 dated 14th September 1944, prepared by G2(I), 1st Airborne Division, 14 September 1944.
37 Harvey, *Arnhem*, 184.
38 Ibid., 184.

Montgomery was aware XXX Corps would face several challenges maintaining its rate of advance. There was the need to move 20,000 vehicles and drive along a narrow road for 65 miles whilst crossing several rivers and canals, with little or no room to manoeuvre off the road if progress was blocked.[39] Ultimately, everything depended on a tight inflexible schedule along a route that favoured enemy defence. A lack of logistical support meant the available transport had to focus on moving supplies, not troops; a tactical consequence of this was a lack of infantry to clear pockets of German resistance that might be encountered.[40] This issue would be compounded if the two flanking British Corps supporting the advance did not keep up, leaving XXX Corps dangerously exposed.

XXX Corps would need to be driven hard, the problem was that its commander, Brian Horrocks was, by this stage, tired and ill, and still suffering from wounds he sustained in 1942; he may not be able to provide the required impetus.[41] (Montgomery would subsequently send him home in December probably because of the state of his mental health).[42]

Montgomery was not alone in harbouring these concerns, doubts were shared more widely. Francis de Guingand, and other members of his staff, were, in fact, against the whole operation. Belchem and Williams, had even tried to dissuade Montgomery from launching the operation, but he would not be put off by this opposition.[43]

Lack of feasibility

Montgomery's immediate staff and indeed those involved in the operation from Dempsey downwards, were opposed to Arnhem as the ultimate objective.[44] Horrocks certainly had doubts about the viability of his Corps' role in the operation; he would prove to be correct.[45] Most commentators agree, arguing that XXX Corps, could have done better, especially the Guards Armoured and 43rd Wessex divisions.[46] Urquhart would accuse XXX Corps of being victory happy by that stage of the war and thus did not push hard

---

39 Clark, *Arnhem*, 101.
40 Harclerode, *Arnhem*, 160.
41 Clark, *Arnhem*, 4.
42 Powell, *The Devil's Birthday*, 14.
43 Bennett, *A Magnificent Disaster*, 221.
44 Buckley & Preston-Hough, *Operation Market Garden*, 212.
45 Powell, *The Devil's Birthday*, 86.
46 Ibid., 239.

enough.[47] The Guards, in the van of the column advancing up the salient in the face of tough opposition were slow, especially after capturing Nijmegen bridge, when they could have pushed on at night, admittedly at some risk to themselves.[48] Buckley argues that it was not just the opposition that was responsible for this; it was also due to the Guards not being suited to their armoured role, lacking the necessary dash and imagination.[49] Bennett goes further, suggesting that the whole military system was flawed and the lack of drive was not necessarily due to individuals.[50] Buckley agrees, concluding the British Army, institutionally, was mentally not capable to making a 65 mile dash into enemy territory.[51] There were certainly occasions when British ground forces were slow in starting different phases of the advance; one unit diary mentioning the fact that 'as usual, the situation eased up in the evening'.[52] Running a critical eye over the situation it is hard to disagree with Harclerode's assertion that the rate of advance was always going to be slow along the single highway,[53] and Powell's conclusion that it is difficult to see how XXX Corps could have managed it.[54]

In summary, several factors would call the viability of Market Garden into question. A lack of supplies and men meant the punch was not as strong as it could have been and probably needed to be. This issue was compounded by proven German ability to rush reinforcements into place and by the possibility, or probability that XXX Corps' rate of advance would be slow. All of this, combined with his staff's opposition to the operation, suggests Montgomery would have assessed Market Garden was, in fact, not entirely feasible. His response to the second question in the DCM process was, therefore, negative. This means that he would have moved into the Defensive Avoidance coping pattern; the final question to explore then, is which form of Defensive Avoidance he adopted.

---

47 Harclerode, *Arnhem*, 171.
48 Ibid., 170.
49 Peaty, *Operation Market Garden*, 66.
50 Bennett, *A Magnificent Disaster*, 194.
51 Buckley & Preston-Hough, *Operation Market Garden*, 209.
52 Bennett, *A Magnificent Disaster*, 193-194.
53 Harclerode, *Arnhem*, 59.
54 Powell, *The Devil's Birthday*, 88.

Bolstering

We have seen that having accepted there was risk to the northern thrust and that Market Garden was a gamble, to the point that there was a big question mark over its feasibility, Montgomery was in the Defensive Avoidance coping pattern. In this state, there are three different forms of coping behaviour, Procrastinating, Scapegoating and Bolstering. Montgomery had a clear view of the way forward and was pushing hard for its adoption, he therefore does not seem to be in the Procrastination coping pattern. He was also pushing Eisenhower to make him overall commander of the land battle and to have sole control of the northern thrust, so he also does not appear to have been looking for the issue to be taken out of his hands (Scapegoating). This just leaves Bolstering. In this mindset he would have been putting a positive spin on the situation, making the best of the available option, the northern thrust / Market Garden; the question is, did he do any of this?

Montgomery's Bolstering

Evidence of Montgomery's Bolstering can be seen in the way that he dealt with the intelligence highlighting risks to the northern thrust that challenged the feasibility of Market Garden. We have already seen a couple of incidents where he was dismissive of this challenging intelligence. When Crown Prince Bernhard met him on 6th September to talk through Dutch underground reports indicating II SS Panzer Corps presence in the area, Montgomery, was dismissive of his concerns.[55] Montgomery also airily waved away the concerns put forward by Bedell-Smith on 16th September, stating that Second Army appreciation's was that the Panzers were not battle worthy.[56] His own Chief of Intelligence also failed to sway him on two separate occasions, 10th and 12th September.[57] The presence of the German forces in the area could not be denied, so it seems that Montgomery's Bolstering behaviour took the form of underestimating their fighting capability and speed of reaction;[58] because, it seems likely, this was a greyer area and had more room for personal judgement to come into play.

---

55 Harclerode, *Arnhem*, 39.
56 Bennett, *A Magnificent Disaster*, 58-59.
57 Harclerode, *Arnhem*, 39.
58 Montgomery, *The Memoirs of Field Marshal Montgomery*, 297.

One practical outcome from this behaviour was that staff officers at Twenty-First Army Group, Second Army and even XXX Corps followed Montgomery's lead and revised down their intelligence appreciations. Indeed, these headquarters would report an improvement in the situation, and that 'previously identified threats had miraculously receded and the only German reinforcements to have appeared in the Low Countries had been put in to thicken up the line they were attempting to form on the Albert Canal'.[59] Furthermore, the FAAA Chief of intelligence was told that there was 'no direct evidence that the area Arnhem-Nijmegen is manned by much more that the considerable flak defences already known to exist.'[60] Second Army, as another example, stated that German units in the area could not refit and prepare a defence at the same time.[61] It seems Montgomery used the intelligence about the likely German resistance to the northern thrust to first cancel Comet and then in reverse to ensure Market Garden went ahead.

The expansion of Comet to Market Garden can also be seen as another form of Bolstering behaviour. Committed to the northern thrust, but faced with growing German resistance in the area, the best way to increase the chance of success was to strengthen the operation by adding in two more airborne divisions and two flanking corps. The alternative would have been to cancel the operation, but this would have meant Montgomery giving up on his war-winning northern thrust strategy, his preferred scheme.

In summary, Montgomery appears to have engaged in Bolstering behaviour. Ignoring and/or watering down the challenging intelligence reports along with expanding and strengthening the operation was the best way to 'make the best of a bad job' and ensure the least-worst option went ahead. He and his staff would have to make the best of what they had.

Conclusion

We have examined Montgomery's thinking at the Decide stage of the OODA Loop through the lens of the Decisional Conflict Model. The various intelligence reports indicating German resistance in the operational area was stiffening posed a serious threat to his northern thrust strategy. The supply and manpower crisis along with a likelihood of rapid German

---

59 Ritchie, *Arnhem*, 148.
60 TNA: AIR 37/217: Information from Northern Group of Armies, Second Army and XXX Corps, as at 1100 hrs, 12th September 1944, by Lieutenant-Colonel A. Tasker, G-2, FAAA, 12 September 1944.
61 TNA: WO 285/3, Second Army Intelligence Summary, 6 September 1944.

reinforcement and the difficulties that XXX Corps would face on the ground meant that Comet/Market Garden was not wholly feasible. This led him into the Defensive Avoidance coping pattern of Bolstering. In this cognitive state, he bolstered or strengthened the least-worst or best available option (in his view) by dismissing the challenging intelligence and expanding the operation. Market Garden represented the closest thing that Montgomery could get to his original notion of a powerful, full-blooded northern thrust. He was, therefore, inclined to take the gamble and carry on regardless of the difficulties; these issues meant there was no margin for error. The major issue then became, as we shall see in the next chapter, was that he did commit errors in the fourth, Act stage of the OODA Loop.

4

Montgomery's Action – Dissonance

'I remain Market Garden's unrepentant advocate'.[1]

Introduction

Montgomery's Tactical Headquarters, Belgium, 25th September 1944

Montgomery's custom was to review his recent operations, so let's join him as he takes stock of Market Garden. British 1st Airborne Division had just been evacuated back across the Rhine, a move that had signalled the end of the advance. The Germans were firmly in control of the Arnhem Road bridge and there was now no hope of an easy bounce across the Rhine. Always keen to identify lessons learned, he was drafting some thoughts on the operation with perhaps one eye on posterity and the almost certain round of blame and accusations that would follow. What should he say? Overall, in his view, the operation had proved 90 per cent successful.[2] It had failed to take the ultimate objective, but XXX Corps had made it 90 per cent of the way. He then began to list the reasons for the (partial) failure of the operation.

Firstly, in his opinion, Eisenhower bore some responsibility. He had failed to provide full logistical support and had allowed Patton's actions to undermine the effort. Secondly, 1st Airborne's landing and drop zones had been too far away from the objectives. He should, perhaps, take some of the blame for this error and have insisted on at least one brigade dropping near to the bridge. Thirdly, there was the disruption caused by the poor weather, but this was just a foible of war. Fourthly, II SS Panzer Corps had badly disrupted

---

1 Montgomery, *The Memoirs of Field Marshal Montgomery*, 298.
2 Powell, *The Devil's Birthday*, 232.

the operation; he had known that the formation was in the Arnhem area, but on reflection, he may have underestimated its fighting ability.[3]

Montgomery's Responsibility

Montgomery's assessment, as laid out later in his memoirs (as listed above), rather neatly pins the blame on Eisenhower, but he himself should also bear a good deal of responsibility for Market Garden going ahead under such challenging circumstances. He was responsible for instigating the operation, wanting to capitalise on the German disorganisation and, as he states, responding to the pressure to clear the V2 launch sites.[4] Ultimately, he could have cancelled the operation; he was certainly not one to squander his men's lives in futile actions, but this was his golden opportunity and he had staked his reputation on it.[5] Market Garden was his operational concept, but stiffening German resistance and lack of full logistical support meant that by the time it was launched, it was probably not a good idea. If this is the case, Montgomery should bear some of the responsibility for the flaws in Market Garden; we therefore need to examine some of the problems with the plan.

Montgomery's Issues

One of the main problems with the planning of the operation was the lack of consultation and coordination at the highest level; the usual planning groups and processes were not followed.[6] Ritchie argues Montgomery wanted to present the airmen with a *fait accompli*; he would later gloat about 'leading air down the garden path'.[7] There was certainly a lack of consultation and liaison, even though Montgomery lacked any real knowledge of airborne operations.[8] Montgomery did not consult Brereton who had previously refused a request by him to use FAAA at Walcheren; Montgomery simply went over Brereton's head to Eisenhower to get Market Garden approved.[9] He was relying on advice from Browning, who, as we shall see, was himself desperate for the operation to go ahead and so was probably not the wisest

---

3 Montgomery, *The Memoirs of Field Marshal Montgomery*, 296-298.
4 Buckingham, *Arnhem 1944*, 72.
5 Powell, *The Devil's Birthday*, 47.
6 Cirillo, *Market Garden and the Strategy of the Northwest Europe Campaign*, 46.
7 Ritchie, *Arnhem*, 206.
8 Ritchie, *Arnhem*, 103.
9 Hamilton, *Monty: The Battles of Field Marshal Bernard Montgomery*, 451.

of counsels.[10] One of the outcomes from this lack of consultation was that Montgomery, Dempsey and Browning all assumed that the air plan for a previous operation, Linnet, (crucially involving two lifts on the first day) could be re-used; the greater distances involved in Market Garden and less available flying light in mid-September meant it could not.[11]

Ritchie argues that the air planners had Market Garden imposed on them giving them only three days to plan it. Subsequent problems with zone selection, the airlift schedule and (lack of) combat support was because the air force was excluded from the planning process. All of this was happening at a time when Montgomery was physically isolating himself at his tactical headquarters. Ritchie concludes that deliberate lack of consultation by Montgomery and Twenty-First Army Group was the main cause for the operation failing.[12]

In summary, there were problems with both the air and (as we saw in the previous chapter) ground sections of the plan, Montgomery was perhaps ignorant of the former but aware of the latter. He seems to have pushed hard for the operation to go ahead, this drive led him to lapse from his usual professional competence in the development of Market Garden. As we have seen, he was faced with trying to solve a difficult problem in stressful and time-pressured circumstances. Research has highlighted the different cognitive biases and errors that individuals can make when making decisions under stressful circumstances.[13] Indeed, stress has been shown to have a number of specific effects on decision-making and information processing, these include: narrowing of timeframes; non-systematic information search; premature closure of the information search process; drawing of inappropriate conclusions from the available information; a reduced ability to consider multiple hypotheses or to integrate information; tunnel vision (an increased focus on central information); and simplifying strategies.[14] The next question we need to address for Montgomery, is what were these lapses and what cognitive biases led him to make them?

---

10 Ritchie, *Arnhem*, 195.
11 Ibid., 119.
12 Ibid., 122.
13 David Hardman, *Judgement and Decision Making: Psychological Perspectives* (Chichester: Blackwell, 2009), 5.
14 Giora Keinan et al, 'Chunking and integration: Effects of stress on the structuring of information', *Cognition & Emotion*, 5, 2, (1987), 133-145.

Montgomery's Lapses

Biases

We shall discuss Montgomery's lapses by exploring the possible issues that can occur during the planning and decision-making process. Each of these issues is linked to or can be caused by a cognitive bias or, in some cases, biases, that we are all prone to making at different times. The first issue to consider is whether Montgomery's focus was too narrow.

Goal Direction

One psychological mechanism that can lead you to adopt a narrow or limited perspective when considering a problem is goal-direction. This occurs when considering an issue with a pre-determined goal in mind, a strong idea of what you want to happen. Your perspective becomes narrowly focused on this goal which means you then do not look more broadly at the topic in hand. The resulting 'tunnel vision' means you may miss vital pieces of information; you may also fail to consider other possible solutions or potential outcomes. This phenomenon was clearly demonstrated by two American psychologists, Chris Chabris and Daniel Simons. In one of their more well-known studies, they asked participants to watch a short video. The video had six people in it, two teams of three. One team of three wore white t-shirts, the other team wore black t-shirts. Each team of three was passing a basketball amongst themselves. The participants' task was to count the number of passes made by the white t-shirt team. This was the 'goal' that the participants were directed towards. At the end of the video, the participants were asked to state the number of passes made by the white t-shirt team, then, crucially if they saw anything of interest. Chabris and Simon found that roughly half the participants missed someone dressed in a gorilla suit walk across the screen, beat his or her chest and then walk off. The gorilla was clearly visible on the screen. The goal of counting the passes meant that the participants' attention was focused on counting the passes and they were essentially 'blind' to the gorilla; Chabris and Simons call this phenomenon Inattentional Blindness.[15] The point is that participants were too goal-directed and didn't look at the bigger picture. The key question here then is, was Montgomery too goal-directed?

---

15 Christopher Chabris & Daniel Simons, *The Invisible Gorilla* (London: Harper Collins, 2011).

Montgomery, as we saw in the first chapter was subject to a lot of situational pressures to act. German resistance appeared to be stiffening and there was pressure to ensure Britain played a key part in Germany's final defeat and maintain its status after the war. He was also under pressure to use FAAA in a strategic role, especially after it had been assigned to him. Montgomery pushed the idea of the northern thrust combined with Market Garden; he therefore gave himself the goal of undertaking the operation.

Chabris and Simons have shown that inattentional blindness or goal-direction works just as well when it is self-imposed. Their original study was conducted in 2010, it has become quite well-known, indeed the video is widely available on the internet. They re-worked the video and produced a second version; the gorilla still makes an appearance but this time the colour of the curtain in the background changes and one of the black t-shirt team drops out of sight. They found that participants watching the video who had seen the original spotted the gorilla but typically missed the other changes; they had essentially given themselves the goal of spotting the gorilla this time round, became goal-directed and demonstrated inattentional blindness. We would argue this is exactly what Montgomery did; he goal-directed himself. Pressured by the situational factors and forced into a Grip reaction in Myers-Briggs terms, he became too focussed on the northern thrust via Arnhem, failing to consider other options or think more broadly about the issues. He also failed to fully appraise the information available to him. There is a second bias that can curtail information search, selective perception.

Selective Perception

Selective perception is a similar mechanism to goal-direction in that it can narrow your focus and limit your information search. The difference is that with goal-direction, you enter the situation with your attention already fixed, you have that pre-determined goal in mind. With selective perception, you enter the situation with a relatively open mind but if something grabs your attention you can become stuck on it, your attention becomes fixed, and your information search becomes too narrow or is cut short. Montgomery's goal-direction meant he was already looking at turning the northern flank of the Siegfried Line. We would argue that selective perception came into play when he identified Arnhem (and not the single crossing at Wesel) as the best option for moving the axis of advance further away from the Americans in the south; this effectively fixed him on this course of action. His information search then became limited and narrowly focused; this can be seen in his lack of consultation with his air force colleagues. It can also be seen in his neglect of the available intelligence, especially Dutch underground reports pointing to threats posed by the German armour, but we will discuss this in more detail later. Essentially, having identified Arnhem as the option that best met his needs, he didn't examine other less risky alternatives. Aiming for Arnhem was a bold imaginative plan, but just how risky was it? The next question to address, is what was Montgomery's risk calculus and was it affected by any cognitive biases?

Framing Effect

As we have seen in the previous chapter, in terms of the Decisional Conflict Model, a key determinant of how you respond to a challenging situation is your perception of risk. Different cognitive biases or psychological mechanisms can affect your risk perception. The way you frame a question or problem, for example, can have a significant effect. More specifically, the way in which you (or someone else) phrase a task or question, can have a dramatic effect on the way you assess risk. Research studies have shown that you are much more likely to be risk averse if a task or question is phrased in positive terms (such as the number of lives that might be saved). In contrast, you are more likely to gamble if the question is phrased in negative terms (such as the potential number of lives lost). One study, for example, presented participants with the problem of dealing with an epidemic that was projected to kill 600 people. The participants were divided into two groups with each group asked to choose between two different vaccine programmes, but being asked with different wording. The outcome of all four programmes were the same (200 lives saved or lost and a 1/3 chance of all 600 being saved or lost) but one set had the question phrased in positive terms, lives saved and thus potential gains, the other set had the question framed in negative terms, deaths caused and thus potential losses.

The experimental effect the researchers examined was the response to the wording, whether it was expressed in positive terms, or negative terms. The study found that when the question was phrased positively (lives saved), 87 per cent of participants were risk averse and opted for the sure gain (of 200 lives saved); whereas when the question was phrased negatively (deaths), only 22 per cent opted for the sure gain, in other words 78 per cent opted to 'gamble' on the probabilistic outcome (where everyone might die).[16] This study, and others, suggest that you are more likely to be risks averse if you frame the situation in terms of potential gains, whereas you are more likely to be risk taking if you frame the situation in terms of potential losses. Risk appetite can, therefore, be determined in part by the way you frame a problem. This cognitive bias is known as the Framing Effect. The Framing Effect, because it involves gains and losses, is partly determined by your goals, what is important to you, what are the specific gains or losses, what is at risk? The questions we need to consider are: what was Montgomery's goal; what did he see as potential gains or losses; and how did this affect his risk appetite?

As we discussed in Chapter One, the strategic debate that occurred in August and September 1944 was between a broad front approach or a single southern or northern thrust. National interests were represented in the two separate thrusts; the southern option would favour Bradley, Patton and the American part of the Allied effort, the northern thrust would favour Montgomery and the British. What was Montgomery's goal in this instance? The functional outcome, the gain, from Market Garden would be ending the war quickly or at least setting the conditions for this. This would be a response to the task pressures he faced. There were also several situational pressures acting on him, especially the Scarcity principle – time was running out. There was also the pressure to ensure that Britain took a leading role in the closing stages of the war. As discussed in previous chapters, he did not appear to be thinking as clearly as he could have been. In Myers-Briggs terms he was in his Grip reaction (and not looking at the situation in his usual realistic/pragmatic manner); and in terms of the DCM, he was in the Bolstering coping strategy. We would argue these factors, along with his rivalry with the Americans, and his growing vanity and egoism, meant that he was more focused on what he would personally lose if Market Garden didn't go ahead. Harclerode is certainly of the opinion that jealousy and personal ambition played a part in Montgomery's thinking.[17]

We would suggest that Montgomery viewed the situation with a negative frame, in terms of what he might lose, and was therefore more risk-taking. This is a key point; the Framing Effect was exerting a strong influence on Montgomery and greatly affected his risk appetite. As many different commentators have argued, Market Garden was a gamble, one

---

16 Daniel Kahneman, *Thinking, Fast and Slow* (London: Penguin, 2011), 363-374.
17 Harclerode, *Arnhem*, 153.

which involved accepting many risks.[18] How did Montgomery deal with the need to accept these risks? As highlighted previously, one method he adopted was to minimise the apparent risk by dismissing the challenging intelligence, especially Dutch underground reports. Fortunately, he also had a readily available solution to hand.

Availability Heuristic

The issue that can arise when you have a readily available solution to a problem is that you may adopt that solution without sufficient thought or critical appraisal. A well-researched cognitive bias called the Availability Heuristic underlies this tendency. The Availability Heuristic is a form of mental shortcut where, when asked to think of something, you recall the answer most readily available to you; it can be most readily available for a variety of reasons: because you have already discussed it at length and thus it is easy to recall; it has been highly publicised in the media, so is easy to picture; or simply because it is vividly dramatic and therefore just easier to imagine. Either way, it is essentially the 'first thing that springs to mind' and you don't look any further. Problems can arise if the most readily available solution is not the right choice.

Numerous research studies have examined the power of the Availability Heuristic. For example, in one study participants believed they were more likely to die in an aeroplane crash or a terrorist attack than of more mundane causes such as diabetes (which results in more fatalities per year).[19] Participants tended to opt for air crashes and terrorist attacks because they are much easier to picture than an invisible health condition. The next question to address here, is whether Montgomery was prey of the Availability Heuristic, was there a ready-made solution to how he was going to make his northern thrust happen? The answer is yes, it was FAAA, the coins burning holes in SHAEF's pockets.

As we have seen, Montgomery was given the use of FAAA on 3rd September and was encouraged to make use of it. He was facing a manpower shortage, and this formation represented the only strategic reserve in the theatre of operations. We would argue that FAAA probably seemed like a perfect and readily available solution to his problem of not only how

---

18 Ritchie, *Arnhem*, 256.
19 Gerd Gigerenzer & Peter Todd, *Simple Heuristics That Make Us Smart*, (Oxford: Oxford University Press, 1999), 56-57.

to provide troops for his northern thrust but also get across the various waterways in Holland. Operation Comet had been driven by Montgomery, with Browning, as we shall see, a willing collaborator, (Sosabowski suggests the pair had developed Comet together on 3rd September).[20] With Browning likely to have been pushing hard to use his airborne forces, FAAA was probably at the forefront of Montgomery's thoughts. Buckingham suggests Browning was involved in the development of Market Garden as well. He argues Montgomery had met with Browning before his fateful meeting with Eisenhower on 10th September, not afterwards as has been generally described.[21] Buckingham proceeds to suggest the plan was promulgated too quickly so must have been developed before the meeting with Eisenhower.[22] As Powell points out, Browning was back in England briefing the outline plan by 1800 the same day.[23] There is good reason to believe, then, that Browning, as the airborne 'expert' and well-known advocate for the capability, had ensured that during early September, FAAA was a readily available solution to Montgomery's problem. The problem was how to utilise this asset to make Market Garden happen? Fortunately, he already had a plan to hand, Operation Comet, expanding this operation would deal with the increased threat and risks. Unfortunately, when developing plans in this way, there is a danger of misapplying experience; this takes us on to our next bias, the Anchoring Effect.

Anchoring Effect

The Anchoring Effect is a cognitive bias that occurs when you use an experience or previous judgement as the starting point when considering a problem, 'what did I do last time?' The problem with this approach is that the prompt can act like an 'anchor' (hence the term Anchoring Effect), and you then fail to adequately adjust your thinking away from this initial assessment. The Anchoring Effect has been examined in a variety of research studies. In an example of this, one half of study participants were first asked to decide whether the percentage of African countries in the United Nations was higher or lower than 10 per cent; the other half were asked to decide whether the percentage was higher or lower than 65 per cent. Both groups were then

---

20 Buckley & Preston-Hough, *Operation Market Garden*, 211.
21 Buckingham, *Arnhem 1944*, 73.
22 Ibid., 74.
23 Powell, *The Devil's Birthday*, 29.

asked to estimate the exact percentage. The first question was designed to act as an anchor; to draw participants to a lower or higher assessment. This is what happened, participants who were given the lower anchor (10 per cent) gave on average a figure of 25 per cent, the higher group (65 per cent) estimated 45 per cent.[24] So, what about Montgomery?

Market Garden was essentially a scaled-up version of Operation Comet. Montgomery took the decision to postpone Comet on 9th September due to the stiffening German resistance that Second Army was facing.[25] The advantage of upscaling Comet was that the existing plans could be recycled. For example, load manifests for gliders assigned to Comet were simply overprinted with Market.[26] The problem in terms of the Anchoring Effect was that the new Market Garden operation, because it was anchored in Comet replicated flaws in the earlier plan: the objectives were still too deep into enemy territory to allow rapid relief of the airborne forces; there was still a lack of suitable landing and drop zones; the airborne forces were still a located too many objectives over too wide an area; the road for XXX Corps was still too narrow to facilitate the rapid advance needed; and numerous water obstacles still needed to be crossed. The irony was that the expansion to three airborne divisions meant only one lift on the first day. This meant that for British 1st Airborne, there was still only one brigade available for the advance to the bridges during the crucial period when speed and surprise was needed; the key point here, is their available combat power was the same as Comet despite the likelihood of increased resistance. It seems the Comet plan was just recycled without sufficient adjustment. The Comet plan acted as an anchor for Montgomery's thinking, and he failed to make sufficient adjustments for the changed circumstances.[27] More careful appraisal may, of course, have highlighted the continuing problems with the operation and maybe kill it off. For it to go ahead, Montgomery needed to maintain his faith in the plan. This is where the next bias comes in, Cognitive Dissonance.

Cognitive Dissonance

Cognitive Dissonance is a psychological bias that invokes a distorted interpretation of information, it occurs when one of your strongly held

---

[24] Scott Plous, *The Psychology of Judgment and Decision Making*, (New York: McGraw-Hill, 1993), 145.
[25] Ritchie, *Arnhem*, 112.
[26] Hamilton, *Monty: The Battles of Field Marshal Bernard Montgomery*, 431.
[27] Ryan, *A Bridge Too Far*, 106.

opinions or beliefs is challenged by incoming information. This creates conflict in your mind; this tension can be resolved in one of two ways. You can either adjust your opinion to accommodate the new information or change in circumstance, which can be taxing both emotionally and cognitively, or you can negate or refute the information in some way so that it does not create or continue to pose a challenge to your opinion. If your opinion is strongly held, the tension is greater, and the dissonance is more acute. If this is the case, the coping mechanisms aimed at dealing with the challenging information (such as denial or questioning the credibility of the source) can be quite extreme in nature to reduce the dissonance.[28] Cognitive Dissonance can thus lead to a less rigorous appraisal of information; the next question then is whether Montgomery engaged in this behaviour?

Montgomery's handling of the intelligence related to Market Garden was discussed in the previous chapter when we examined his perception of risk in terms of the Decisional Conflict Model. As we saw, the intelligence picture was building towards indicating greater German strength, especially in the Arnhem area. Montgomery's Chief of Intelligence, Brigadier Bill Williams, alarmed by Ultra decrypts, tried to persuade his boss on both 10th and 12th of September of the presence of II SS Panzer Corps. The Dutch Crown Prince Bernhard had, on 6th September, already laid out the case, using underground reports, for the presence of this formation and the threat it posed to the operation. On 15th September, Kenneth Strong, Eisenhower's Head of Intelligence and his Chief of Staff, Walter Bedell-Smith, both flew to visit Montgomery to warn him about the threat.[29] There is, therefore, sufficient evidence to suggest that Montgomery was receiving information that challenged his strongly held opinion, that Market Garden routed through Arnhem was the best strategy for the Allies to adopt; this is the first condition for Cognitive Dissonance. It seems he also engaged in behaviour that conformed to the second component, actively negating the challenging information. Montgomery effectively ignored the Ultra intercepts when these were presented to him.[30] He also dismissed Prince Bernhard's information, belittling the intelligence partly due to it being based on Dutch underground reports, the same movement that had been penetrated by the Germans earlier in the war.[31] Finally, Montgomery is reported to have 'airily waved away'

---

28 Plous, *The Psychology of Judgment and Decision Making*, 22-30.
29 Harclerode, *Arnhem*, 39.
30 Irving, *The War Between the Generals*, 281.
31 Ryan, *A Bridge Too Far*, 86.

Strong and Bedell-Smith's concerns; he was dismissive of the intelligence, adding with two days to go it was too late to make changes anyway.[32] This all suggests that Montgomery was experiencing Cognitive Dissonance in the run up to Market Garden so he could maintain confidence in his planned operation. The problem is that we can be too confident in the likely success of something, which brings us on to the penultimate issue to consider, Optimism Bias.

Optimism Bias

If you are estimating the outcome of a future event, the chances are that you are very likely to be overconfident in your expectation of success. This is driven by your innate Optimism Bias; your tendency to believe that you are less likely than is the case to experience a negative event and more likely to experience success. This tendency is further compounded by the fact that your confidence in your judgements does not tend to be associated with your actual accuracy; we remain confident despite previous feedback to the contrary.[33] In planning, it means that you are likely to be overly positive or confident in your belief that your plan will work.[34] Just think of IT or large scale construction projects that have taken much longer and proved more expensive to build than the original estimate. The next question we need to address in this chapter is, therefore, whether Montgomery was overconfident in his plan, was he suffering from Optimism Bias?

We have already examined the victory euphoria existing in most parts of SHAEF by September 1944. This is exemplified in the SHAEF intelligence summary for 26th August that we have already seen: 'the August battles have done it; brought the end of the war in Europe in sight, almost within reach'.[35] Irving suggests that Montgomery was not immune to this feeling.[36] There appears to have been every reason for him to adopt a hopeful outlook, and that Montgomery was indeed suffering from Optimism Bias and was thus overconfident in Market Garden's chances of success. Ryan agrees, pointing out Montgomery thought that the Germans would 'crumble'.[37] This second point, the reaction of German forces, speaks to the last bias that we shall

---

32 Ritchie, *Arnhem*, 131.
33 Hardman, *Judgement and Decision Making*, 94-103.
34 Ibid., 104.
35 TNA, WO 219/1922, SHAEF Intelligence Summary No.23, 26th August 1944.
36 Irving, *The War Between the Generals*, 249.
37 Ryan, *A Bridge Too Far*, 95.

consider in this chapter, a failure to consider external factors that might affect a plan.

Planning Fallacy

The Planning Fallacy is similar in several ways to Optimism Bias in that it refers to a tendency for you to be overly ambitious when planning something. Strictly speaking, the Planning Fallacy involves the underestimation of the time required to complete a future task. This is partly based on optimism but also due to failing to consider the external factors that may potentially delay or disrupt the task; focussing only on internal factors to the exclusion of external influences, especially chance factors.[38] The result of this bias is taking a narrow, inwardly focussed approach to planning and a neglect of external considerations. The impact is on not just the (extra) time needed or the costs incurred, but also on the likelihood of a successful outcome. The final question to consider in this chapter, then, is the extent to which Montgomery may have focused on internal planning considerations and failed to factor in external influences. The two obvious neglected influences are the German resistance and the weather, both of which, as we saw at the start of this chapter, he points to, with hindsight, as reasons for the failure of Market Garden.

Several individuals involved in the operation have remarked on the nature of the German reaction to Market Garden. Urquhart states in his memoirs the German reaction to the airborne landings was 'impressive'.[39] Gavin, who commanded the US 82nd Airborne Division at Nijmegen agrees, adding that the air of optimism meant that there was little concern over the likely German resistance.[40] Montgomery admits in his memoirs that he was aware that II SS Panzer Corps was in the Arnhem area, but that he underestimated its fighting ability.[41] Montgomery was also guilty of overestimating the airborne forces and XXX Corps' fighting ability.[42] These two points in combination constitute the Planning Fallacy in action, a focus on internal considerations to the exclusion of external factors.

---

38 Hardman, *Judgement and Decision Making*, 108.
39 Urquhart, *Arnhem*, 202.
40 James Gavin, *On to Berlin: Battles of an Airborne Commander 1943-1946* (New York: Viking Press, 1978), 161.
41 Ritchie, *Arnhem*, 148.
42 Ibid., 143.

Montgomery also blamed the weather for the failure of the operation;[43] his Chief of Staff, Freddie de Guingand, agrees and has stated that with better luck with the weather the Market Garden might have succeeded.[44] The weather did impact on the operation, fog in England on the third day meant the third drop was delayed, and resupply was disrupted.[45] Powell, accepts that bad weather played a part but argues that it did not have a material impact on the outcome of the operation.[46] Ultimately, the weather did not affect operations on the crucial first day of the operation but it did have an impact on later airborne drops and helped the Wehrmacht to win the battle of reinforcements as the operation dragged on.[47]

In summary, the weather and degree of German resistance were two external considerations Montgomery did not sufficiently factor into his plan. These two issues were amongst the reasons he listed in his memoirs for operation failing. It seems clear, then, that Montgomery suffered from the last cognitive bias we have considered, the Planning Fallacy.

Conclusion

This chapter has discussed several cognitive biases that appear to have been prevalent in Montgomery's thinking during his planning for Operation Market Garden. The need to get agreement on the northern thrust meant he adopted a too narrow view of the available options and when he alighted on the neat solution of using FAAA, he stuck too rigidly to his guns. This narrow focus meant he failed to factor in external factors and was too confidently focused on internal planning considerations. As Montgomery, sitting at the strategic level, was responsible for instigating Market Garden and developing the operational concept, perhaps the most significant error was his narrow view on the options available to him to further prosecute the war in late August/early September. In the final chapter of this part of the book we will examine how a Structured Analytical Technique, the Analysis of Competing Hypotheses, might have helped him to be more critical in his thinking.

---

43 Powell, *The Devil's Birthday*, 242.
44 Clark, *Arnhem*, 334.
45 Harclerode, *Arnhem*, 163.
46 Powell, *The Devil's Birthday*, 242.
47 Bennett, *A Magnificent Disaster*, 198.

5

Dilemma

Introduction

In the final chapter in this section, we examine how some of the biases described earlier could have been mitigated. We will use a Structured Analytical Technique to examine the strategic dilemma that faced the Allies in late August and early September 1944, namely, how to proceed now that the advance was slowing. Using a technique called the Analysis of Competing Hypotheses to apply Critical Thinking to the problem, we explore which of the available strategies was the best option. Before we do this, it will be useful to summarise again the situation faced by the Allies.

Situation

During August there was a strong sense of victory euphoria in the Allied camp; the prevailing view was that the Germans were essentially beaten. The daily intelligence summary of Second Army for 5th September is a good example of this optimistic assessment: 'The August battles have done it; brought the end of the war in Europe in sight, almost within reach'. By early September, however, there were signs that German resistance was stiffening across the front. Within a short space of time of Second Army resuming the offensive on 6th September, it was clear that the German defences had been reinforced and strengthened.[1] The key question, in the face of this renewed resistance, was how best to move forward? This is where we can turn to the Analysis of Competing Hypotheses technique to help.

---

[1] Hibbert, *The Battle of Arnhem*, 22.

	Hypothesis 1	Hypothesis 2	Hypothesis 3	Hypothesis 4
Criterion 1				
Criterion 2				
Criterion 3				

Table 5.1   ACH Matrix Template

Analysis of Competing Hypotheses

This technique is a useful tool originally designed to examine a problem where there are several hypotheses that could explain either the cause of the event, its nature or how it might unfold in the future. Possibly all three. The process involves taking each (causal) hypothesis and analysing information that is either inconsistent or consistent with it and rejecting hypotheses that do not stand up to scrutiny.[2]

The process for this technique follows a series of analytical steps. Firstly, an analysis of the situation to identify, the hypotheses in play. The hypotheses should be mutually exclusive; if one hypothesis is true, the others must be false. The next step is to generate a list of critical or significant pieces of information that relate to the event, these are the significant points of evidence or arguments involved. These are the criteria used to assess the hypotheses. The third step is to create a table or matrix with the hypotheses as the column headings and the relevant information points as the rows. An example is shown in Table 5.1.

The next step is to analyse each information criteria in terms of whether it is either consistent or inconsistent with each hypothesis, or indeed not applicable or not relevant. Finally, the matrix is completed by putting '1' in the appropriate cell if the information is consistent or '0' if it is inconsistent with the hypothesis in question. The hypothesis with the most '1's is the most viable option.

The technique can also be used to select different possible courses of action to solve a problem. This is what we will use it for here. In our case, the hypotheses are the different courses of action that were available to the Allies, and the information criteria are the requirements that needed to be met by the selected option. The next question is to examine what were the requirements?

---

[2]   Richards Heuer & Randolph Pherson, *Structured Analytical Techniques for Intelligence Analysis*, (Los Angeles: Sage, 2015), 181-192.

What objectives needed to be achieved and what were the factors impacting on the planning considerations?

Requirements

Clear the Scheldt Estuary
As we have seen, there was an urgent need to resolve the Allied supply problem. British and American divisions needed 700 tonnes a day to operate but the front-line formations were 300 miles from Normandy. The strategically vital port of Antwerp was seized on 4th September with its facilities virtually intact. The port had an unloading capacity of 40,000 tonnes per day and was much closer to the front line and Germany. The port facilities were, however, 60 miles from the sea and ships would be required to navigate the Scheldt Estuary. The problem was that German forces controlled the southern and northern banks (Walcheren Island) of the estuary.[3] This is why Eisenhower told Montgomery on 10th September to first establish the Arnhem bridgehead and then clear the Scheldt Estuary with his whole force. Montgomery considered using airborne troops to seize Walcheren Island but was told by FAAA that the area was not suitable for airborne troops.[4] There was clearly, therefore, a need to prioritise opening Antwerp by clearing the Scheldt Estuary.

Maintain pressure
Another consideration was the need to maintain momentum, to keep up pressure on the retreating Wehrmacht to take advantage of German dislocation. Resistance was stiffening by the day, and any pause would allow the Wehrmacht more time to reorganise and reconstitute itself. Therefore, whichever solution was adopted, there was a need to maintain the Allied advance and keep up pressure on the German Army before it could stabilise its various frontlines.

Bring war to a close quickly
A key consideration was the broader situation in Europe. With the Red Army advancing rapidly in the East, any delay in Western Europe would mean more territory would be ceded to Stalin. The western Allies therefore

---

3 Harclerode, *Arnhem*, 21.
4 Ritchie, *Arnhem*, 115.

needed to bring the war to close as rapidly as possible to avoid future Soviet occupation of Eastern Europe.

Conserve manpower
Chapter One outlined the manpower crisis facing the western Allies and Britain in particular. The chosen course of action would need to be frugal in its use of manpower. An attritional campaign would certainly not achieve this whereas a brief operation that brought quick results would. Although a brief operation might suffer more casualties in the short-term, it would more likely be sparing of lives in the longer-term.

Frugal with resources
On a similar theme, without the port of Antwerp being open, the selected operation would need to be frugal in its use of resources. As with the manpower issue, a large scale or protracted campaign would certainly not achieve frugality, whereas a focused operation that brought quick results would use less supplies.

Politically neutral
As explained, a big headache for Eisenhower was the need to keep the Allies together and the potential political ramifications of any decision that he took. There was fierce rivalry between the British and American Army Groups, with both commanders lobbying hard for preference in terms of troops and supplies. Eisenhower was cognisant of the need to stay as non-partisan as possible.[5]

Utilise First Allied Airborne Army in a strategic role
Finally, although of lesser importance, another factor in the decision calculus was the need to utilise FAAA in a strategic role.[6] This has been discussed at length in previous chapters, it was 'coins burning a hole in SHAEF's pocket'. The selected option, would, therefore, if it did not exactly need to use FAAA, would certainly benefit from doing so. These requirements can be put into the matrix as the 'information criteria'; this is shown in Table 5.2.

These are the factors that we can use to compare the hypotheses. The next step is to identify each of the different hypotheses or possible courses of action.

---

5 Ritchie, *Arnhem*, 90.
6 Clark, *Arnhem*, 96.

Requirements	Hypothesis	Hypothesis	Hypothesis	Hypothesis	Hypothesis
Open Antwerp.					
Keep pressure on retreating Germans.					
Bring the war to a close quickly.					
Conserve manpower.					
Be frugal with resources.					
Be politically neutral.					
Use FAAA strategically.					

Table 5.2 ACH Matrix – Criteria

Options

Broad front strategy (Hypothesis One)
The broad front strategy, favoured by Eisenhower, saw the Allies maintaining their advance across a wide front, or at least utilising at least two thrusts, one either side of the Ardennes. Eisenhower favoured occupying both of Germany's key industrial heartlands, the Saar in the south and the Ruhr in the north and the broad front strategy did that. It also had the advantage of not favouring either Army Group.

Southern thrust (Hypothesis Two)
A single thrust to the south. Bradley, as we have seen, initially favoured an attack with an airborne drop in the Aachen area. Bradley told Montgomery on 3rd September that he preferred an operation even further south, between Koblenz and Mannheim.[7] This operation would have the advantage of targeting the Saar industrial region but, it would only be feasible if Montgomery's forces were halted in the north.

Northern thrust (Hypothesis Three)
Essentially a mirror image of its southerly neighbour; the problem was the

---

7 Harclerode, *Arnhem*, 214.

Requirements	Broad Front	Southern Thrust	Northern Thrust	Scheldt Estuary	Strategic Bombing
Open Antwerp.					
Keep pressure on retreating Germans.					
Bring the war to a close quickly.					
Conserve manpower.					
Be frugal with resources.					
Be politically neutral.					
Use FAAA strategically.					

Table 5.3 ACH Matrix – Criteria and Hypotheses

reverse. It would have the advantage of targeting an industrial region, the Ruhr, but again, it would only be feasible if Bradley's advance was halted in the south.

Clear Scheldt Estuary (Hypothesis Four)
An operation to clear the Scheldt Estuary and open the port of Antwerp, as Eisenhower had ordered Montgomery to do.[8]

Strategic bombing (Hypothesis Five)
A final, somewhat wildcard option was to rely on the strategic bombing campaign to bring Germany to its knees. The campaign was successfully choking off petrol supplies by air bombardment and inflicting considerable damage in other areas. The problem was that it would not bring the war to a close by occupying Germany, which, the Allies having agreed on unconditional surrender, they needed to achieve.

These, then are the strategic options that can be dropped into the ACH matrix as the 'hypotheses'; this is shown in Table 5.3.

The final step is to evaluate each strategic option confronting the Allies against the decision criteria or requirements. In terms of opening Antwerp, only the Scheldt Estuary option would achieve this aim. The broad front

---
8 Buckingham, *Arnhem 1944*, 71.

Requirements	Broad Front	Southern Thrust	Northern Thrust	Scheldt Estuary	Strategic Bombing
Open Antwerp.	0	0	0	1	0
Keep pressure on retreating Germans.	1	1	1	0	0
Bring the war to a close quickly.	0	1	1	0	0
Conserve manpower.	0	1	1	0	1
Be frugal with resources.	0	1	1	1	1
Be politically neutral.	1	0	0	0	1
Use FAAA strategically.	0	1	1	0	0
Total	2	5	5	2	3

Table 5.4 ACH Matrix – Complete

approach and both the single thrusts, southern and northern, would maintain pressure on the retreating Germans. The broad front approach, Scheldt operation and, arguably, the strategic bombing campaign, would not bring the war to a close quickly: the two single thrusts, if successful, stood a chance of achieving quicker victory and would, in theory, be more conservative in terms of manpower as they would avoid a broader attritional battle. The strategic bombing campaign would also incur less casualties due to the more limited numbers of air force personnel involved. The Scheldt operation would have to be followed by another operation to continue the advance and so would not necessarily be conservative in terms of casualties in the long run. In terms of frugality with supplies, the broad front strategy, due to its large scale, protracted and attritional nature, would not meet this criterion, whereas the others would. The broad front strategy and strategic bombing campaign would be politically neutral, the other three options would favour either the British or the Americans. Finally, only the two single thrusts would utilise FAAA in a strategic role (as we have seen, it was an option for use on the Scheldt operation, but FAAA argued it was not a strategic use of itself).

To make the comparison of options we can enter these assessments into the matrix with '0' being awarded if the option does not meet the requirement and '1' if it does. The different options can then to summed to provide an overall 'total' score; this is shown in Table 5.4.

Requirements	Broad Front	Southern Thrust	Northern Thrust	Scheldt Estuary	Strategic Bombing
Open Antwerp.	0	0	0	1	0
Keep pressure on retreating Germans.	1	1	1	0	0
Bring the war to a close quickly.	0	1	1	0	0
Conserve manpower.	0	1	1	0	1
Be frugal with resources.	0	1	1	1	1
Be politically neutral.	1	0	0	0	1
Use FAAA strategically.	0	1	1	0	0
Clear V2 launch sites	0	0	1	1	0
Total	2	5	6	3	3

Table 5.5   ACH Matrix – Revised

As can be seen from the table, the two single thrusts come out as the best options with this scoring approach. It is possible to award more than a simple score of 1 and give ratings of 2, 3, 5 or even 10 where an option meets the requirement to a greater or lesser extent. The requirements can also be weighted in terms of importance. For example, opening Antwerp could be weighted double which would raise the Scheldt operation up in the scoring, however, trebling the weighting for the need to keep pressure on the Germans and to finish the war quickly, would make the two single trust options clearer front runners. Using graduated scoring and weighting both require a robust justification based on more detailed analysis than we have space for here, so, for the purposes of this discussion, we will keep the simple scoring approach. This still leaves us with the problem of which of the two single thrust options to choose. As discussed in previous chapters, V2 weapons began to fall on London (and Brussels) in early September. SHAEF was put under intense political pressure to move quickly to clear the V2 launch sites from the Antwerp-Utrecht-Rotterdam area. Adding this requirement into the matrix changes the calculus as the northern thrust strategy and Scheldt operation are the only two options that meet this criterion; this additional analysis is shown in Table 5.5.

The outcome of adding this requirement to the matrix is to push the northern thrust strategy slightly ahead in the scoring. This analysis also suggests that the decision to go with the northern thrust held a large political component, rather just being based on military considerations. Indeed, Montgomery claims he used the message detailing the V2 attacks as justification for opting for the more northerly Arnhem axis over the more southerly Wesel line of advance.

Different requirements and different approaches to the scoring would of course potentially lead to different outcomes, but based on this analysis, Operation Market Garden, somewhat controversially, is the preferred option, at least at the level of the strategic debate. It was worth perhaps a gamble as some commentators have suggested.[9] To be worth the gamble, it needed some chance of success, the next question to be addressed is how feasible was it? To do this, we need to drop down to the operational level and examine key issues affecting the level of risk and the measures taken by the individuals that addressed those issues, especially the commander of the operation, Lieutenant-General Frederick 'Boy' Browning.

---

9 Powell, *The Devil's Birthday*, 243.

Part Two

Reproduced with kind permission of the Imperial War Museum.

Browning

6

Browning's Observation – Consistency

'Coins burning holes in SHAEF's pocket'.[1]

Introduction

Methodological Issues

The first chapter in this section summarises Browning's background and service in the war along with his involvement in the development of and advocacy for British Airborne Forces. Before we get into this discussion it is worth pausing to reflect on some methodological issues because Browning is somewhat of an enigmatic character. He did not keep a diary that we can access and few of his letters survive. Of the letters that do remain, most are to his wife (the novelist and socialite), Daphne du Maurier, and are partly written in a personal code. Browning has only one biographer, Richard Mead, who although not an apologist, is somewhat sympathetic to his subject; we need to be cautious using his work.

Browning has also attracted a lot of criticism, especially his role in Market Garden; some commentators thus have a rather jaundiced view of him. The film *A Bridge Too Far* certainly portrays Browning as the villain of the piece; the scene where Browning (played by Dirk Bogarde) dismisses the concerns of his intelligence officer, exemplifies this. We need, therefore, to bear all of this in mind when examining Browning's thinking and actions; let's join him at his headquarters just before the start of the operation.

---

[1] Bennett, *A Magnificent Disaster*, xv.

Browning's Headquarters, Moor Park, England, 16th September 1944

There was less than 24 hours to go before the launch of Market Garden, the biggest airborne operation in history, and Lieutenant-General Frederick 'Boy' Browning was in command of it. He was something of a glamorous figure, good-looking, charming and, being a Guards officer, always immaculately turned out. He also enjoyed some notoriety, being the husband of a famous author. He had enjoyed a long career in the British Army and for the last couple of years had been intimately involved in the development of Britain's Airborne Forces. He was under a lot of pressure to make a success of the operation and given his background he was perhaps best placed in the British Army to do this.

Background

Browning attended West Downs school until the age of 14, where he apparently did not distinguish himself academically, moving on to Eton in September 1910 where he 'did not have a notably sparkling career'. His achievements at Eton were 'scholastically average' and he joined the Army Class in 1913, a move for those who were generally not strong academically. Browning failed the entrance exam for Sandhurst in November 1914, doing particularly poorly in Mathematics. He was, however, recommended by his headmaster at Eton who was able to get him in using a provision in the Regulations for Admission to the Royal Military College. The picture that emerges of Browning intellectually is not strong.[2]

In terms of military service, Browning was a Grenadier Guards officer and was very much imbued with the Guards traditions and values. His service in the First World War, although he missed the worst of the bloody battles through a combination of luck and timing, is described as a 'good war',[3]. He did particularly well in one engagement earning himself a reputation as a 'fighting soldier'.[4] Browning took part in the final Allied offensive in 1918, advancing rapidly across Belgium; Mead suggests this experience affected his thinking when involved in the similar pursuit, of which Market Garden was a part, in 1944. Browning made steady progress during the inter-war years. He did not attend staff college, however, which may have impacted on

---

[2] Mead, *General 'Boy'*, 8.
[3] Ibid., 28.
[4] Mead, *General 'Boy'*, 20-23.

his planning abilities, which Bennett describes as 'sketchy and rigid' adding he 'lacked the imagination and practical ability to modify a plan'.[5]

Regarding his personality, Browning has been described as 'hard to estimate'.[6] Roy Urquhart certainly found him somewhat reserved and hard to get to know.[7] As with most people, there are conflicting descriptions. Brian Urquhart, his intelligence officer, described him as 'sympathetic and tolerant',[8] that his nickname 'Boy' was accurate as he had never really grown up.[9] Members of his staff have, at different times, admired him.[10]

Conversely, he is regularly described in other quarters as: 'high handed and arrogant';[11] 'overbearing and arrogant' (again);[12] and 'vain and aloof'.[13] It appears that he was particularly patronising towards his American colleagues, which as we shall see, caused some problems during the run up to Market Garden; he was disliked by them.[14] Whilst 'naturally charming and highly sociable, with a pronounced sense of humour', he could also be an 'explosively-tempered disciplinarian'.[15] This disciplinarian trait probably has something to do with his Guards background, (the disciplinarian bit not the temper); he was certainly always immaculately turned out. Several commentators have also described Browning as highly ambitious, (socially as well as militarily).[16] We shall examine this more closely in the next chapter. His motivation is one of the key questions that this part of the book seeks to address; suffice to say, his ambition played a key role in his pushing for Market Garden to go ahead.

In summary, although a somewhat mixed picture emerges, the over-riding description of Browning seems to be a charming fighting soldier but one who was arrogant, patronising and ambitious. This ambition was not supported by a strong intellect or an aptitude for staff work. We shall see how these characteristics played out, especially in terms of how he responded to the situational and task pressures he faced.

---

5 Bennett, *A Magnificent Disaster*, 13.
6 Ibid., 12.
7 Baynes, *Urquhart of Arnhem*, 70.
8 Bennett, *A Magnificent Disaster*, 12.
9 Mead, *General 'Boy'*, 14.
10 Ibid., ix.
11 Buckingham, *Arnhem 1944*, 17.
12 Powell, *The Devil's Birthday*, 38.
13 Clark, *Arnhem*, 67.
14 Powell, *The Devil's Birthday*, 38.
15 Mead, *General 'Boy'* 42.
16 Clark, *Arnhem*, 67.

Situational Pressures

Browning was appointed to command Market Garden, therefore responsible for an operation that was challenging, complex, and had to be mounted in a short timescale. There were, therefore, several pressures he faced that were due to the task itself. As with Montgomery, there were also several situational factors adding to the pressure, all of which would have encouraged Browning to proceed with the operation. We can again use Cialdini's framework of the principles of persuasion (Authority, Conformity, Scarcity and so on) to explore these in more detail. Let's return to Browning's Headquarters to put ourselves in his shoes and look at the situation from his perspective.

Browning's Headquarters, Moor Park, England, 16th September 1944

Even if he had been so inclined, there was little Browning could do to challenge the situation he was facing. At the most basic level, he was in receipt of a lawful order; he was subject, therefore, in large degree to explicit authority. To have refused the order would have meant challenging his chain of command and essentially disobeying orders. Not only was he under orders, but he was also subject to a good deal of implicit authority. The problem was that he was also on something of a sticky wicket as the formation of FAAA had witnessed quite a power struggle.

Politics

The American general, Lewis Brereton had been appointed to command FAAA when it was formed in the summer of 1944.[17] Browning was unhappy with this appointment, believing he was slightly more senior than Brereton and knew more about airborne operations, British ones at least. Browning had instead been made deputy commander whilst also remaining as General Officer Commanding (GOC) British I Airborne Corps.[18]

Brereton's appointment to command FAAA made sense as he was an airman. The main challenge for airborne operations, identified from previous experience, was the accuracy of the airlift.[19] Some commentators have

---
[17] T.B.H. Otway, *Airborne Forces* (War Office official monograph, 1951), 201-206.
[18] Mead, *General 'Boy'*, 108.
[19] Ritchie, *Arnhem*, 90.

suggested he was, however, a poor choice to meet this and other challenges.[20] He had previously commanded US Far East Airforce in the Philippines ahead of the Japanese invasion in December 1941, where his indecision over the intelligence he received resulted in nearly all US planes being destroyed on the ground. He was certainly a difficult character and could be intransigent at times.[21] Eisenhower told Brereton to be bold and imaginative when he was appointed, but Powell suggests Brereton was a little bemused by his job and didn't meet this challenge.[22] The British were certainly suspicious of his appointment, viewing FAAA Headquarters as an American institution. Conversely, Ridgway and Gavin were worried the formation of FAAA meant unnecessary British control, particularly that of Browning.[23] The command appointment of FAAA was ultimately a poor compromise; Brereton did not want the job, considering it a demotion whilst Browning was unhappy as he wanted the top seat for himself.[24] This situation was compounded by the fact that the pair had little in common and did not get on with each other.[25]

There is a suggestion the American contingent had blocked Browning's appointment to command the formation.[26] His fellow American officers disliked Browning on a personal level. He has been accused of being overbearing, arrogant and patronising.[27] Matthew Ridgway, his nearest rival, was certainly highly suspicious of him.[28] Browning had gotten off to a bad start with Ridgway when he had argued with him over the allocation of aircraft between British and American divisions for an earlier operation; this had been compounded by Browning organising an inspection of one of Ridgway's battalions without asking or even informing him.[29] Ridgway had been disappointed when Browning was appointed to command Market Garden and appeared to be waiting in the wings to replace him.[30] This possibility had been made more likely by Browning's behaviour over a dispute during the planning of a previous operation.

---

20 Bennett, *A Magnificent Disaster*, 32.
21 Harclerode, *Arnhem*, 164.
22 Powell, *The Devil's Birthday*, 234.
23 Otway, *Airborne Forces*, 201-206.
24 Buckingham, *Arnhem 1944*, 63.
25 Powell, *The Devil's Birthday*, 38.
26 Ritchie, *Arnhem*, 90.
27 Powell, *The Devil's Birthday*, 38.
28 Mead, *General 'Boy'*, 109.
29 Ibid., 90.
30 Ibid., 118.

Browning had threatened to resign as deputy commander over the lack of issue of maps for a previously cancelled operation, Linnet II.[31] Both Buckingham and Ritchie query the reason for the threatened resignation, suggesting instead it was really an attempt to kill off the operation as it was due to take place in the US sector, and thus benefit the Americans, not Montgomery.[32] Brereton had called Browning's bluff, saying he would forward the resignation letter to Eisenhower and replace Browning with Ridgway.[33] Browning, faced with this threat, had withdrawn his resignation. He was now in a difficult position with Brereton.[34]

Weak Position

Browning was on thin ice. His demeanour, his American colleagues' mistrust, the underlying politics in FAAA and his resignation threat had all undermined his position, making him vulnerable to implicit authority pressures and thus, along with the explicit authority of a lawful order, made it very difficult for him to push back on Market Garden. Browning, however, was not particularly inclined to push back for he was keen to see Market Garden go ahead as it would use FAAA in a truly strategic role. This desire can be best seen through the lens of another of Cialdini's principles, Commitment and Consistency.

Commitment and Consistency

The principle of Commitment and Consistency refers to situations where if you commit to something such as a promise, an idea or a goal, you are more likely to honour that commitment because you want to maintain consistency between your thoughts and actions. New Year's resolutions are an obvious example of this, as are pledges made in self-help groups such as Alcoholics Anonymous. The power of this principle is that breaking the commitment would be incongruent with your self-image and that we do not like a discontinuity between our attitudes and actions.[35]

---

31 Powell, *The Devil's Birthday*, 39.
32 Ritchie, *Arnhem*, 100.
33 Mead, *General 'Boy'*, 110.
34 Harvey, *Arnhem*, 39.
35 Gregory Maio & Geoffrey Haddock, *The Psychology of Attitudes & Attitude Change* (London: Sage, 2015), 73-74.

Browning's Commitment

Browning could look back over the last few years with pride. He and the British Airborne Forces had come a long way. By September 1944, the Allies and Browning himself, had invested a lot of time, effort and money developing the airborne concept.

The British Prime Minister, Winston Churchill, had been keen to develop an airborne capability after watching German success in the invasion of Norway in April 1940, the Low Countries in May of the same year and Crete twelve months later.[36] A parachute training school had been established at Ringway in Manchester with No.2 Commando reassigned to become airborne.[37] Further developments saw the establishment of 1st Parachute Brigade under Brigadier Richard 'Windy' Gale in September 1941; the Brigade included such figures as Eric Down and Gerald Lathbury, both of whom we shall encounter later. This was followed the next month by the establishment of 1st Airlanding Brigade Group under Brigadier George 'Hoppy' Hopkinson with Browning appointed Commander Paratroops and Airborne Division in the following month. Browning was thus involved in developing the airborne concept at an early stage, albeit not at the start. Although incorrectly referred to as the 'Father of the Airborne' he had worked hard to expand Britain's airborne capability using his connections and political savvy to ensure its survival. This had been essential as there had been a political battle with the RAF from the outset.

The RAF had attempted to strangle the airborne concept at birth; the key issue being the drain on available aircraft and distraction from the strategic bombing campaign. Browning had been able to secure the support of the Chief of the Imperial General Staff (CIGS), Alan Brooke, during his struggle to establish and then expand the airborne forces. Ultimately, the RAF had, after much wrangling, obtained control of the planning of the air side as the price for its subsequent cooperation.[38] The capability had been used in North Africa in 1942 and then in Sicily in 1943, with decidedly mixed results. Three divisions, British 6th Airborne and the US 101st and 82nd, were used on D-Day, again with mixed results, but by the summer of 1944, the airborne concept had been firmly established. This had been underscored by the formation of FAAA on 8th August at great expense and there was now a

---

36 Mead, *General 'Boy'*, 66.
37 Ibid., 67.
38 Buckingham, *Arnhem 1944*, 12-13.

lot of pressure to use it before the war ended; deploying it in Market Garden would therefore be consistent with this commitment of resources.

First Allied Airborne Army enjoyed a powerful Order of Battle; its combat capability was made up of two corps, one American and one British. The US XVIII Corps, commanded by Matthew Ridgway, comprised 17th, 82nd and 101st Airborne divisions; the latter two were slated for use in Market Garden under James Gavin and Maxwell Taylor respectively. The British I Airborne Corps was comprised 1st and 6th Airborne divisions under Roy Urquhart and Richard Gale, with the former jumping at Arnhem. It also had the 1st Polish Independent Parachute Brigade under Stanislaw Sosabowski and the British air portable 52nd Lowland Division, commanded by Edmund Hakewill-Smith. The airlift capability of FAAA comprised IX US Troop Carrier Command under Paul Williams and RAF 38 and 46 Groups, commanded by Leslie Hollinghurst.[39] First Allied Airborne Army now formed SHAEF's only strategic reserve and was something of a political football, it was also not without its critics.

Coins burning holes in SHAEF's pocket

The airborne forces had been growing since 1941 and had been steadily gaining momentum during 1943 and especially 1944; the capability was getting bigger and bigger. It was the sheer size of the effort that was attracting criticism.[40] As we have seen, FAAA were 'coins burning holes in SHAEF's pocket'.

The formation comprised one sixth of the manpower available to Eisenhower and his only strategic reserve; he was coming under increasing pressure to use it.[41] The Allies had ten airborne divisions in existence, a considerable force which even Matthew Ridgway described as a fashionable new toy.[42] By August, the airborne community was in the doldrums with a growing feeling that the effort had not been worth it, and that airborne operations were simply a bonus.[43] Eisenhower was being pressured by George Marshall and Hap Arnold back in Washington and so wanted

---

39 Clark, *Arnhem*, 96.
40 Powell, *The Devil's Birthday*, 246.
41 Ibid., 11.
42 Ibid., 249.
43 Ibid., 14.

FAAA to devise a daring and imaginative operation to justify all the effort.[44] Eisenhower assigned the formation to Montgomery at Twenty-First Army Group for his northern thrust and urged him to use it.[45] There was, therefore, ultimately a lot of external pressure to see the investment pay dividends; Browning was intimately involved in all of this.

In summary, Browning had been involved in developing British Airborne Forces and wanted to see the concept fully used. Whilst not quite the 'father of the airborne', he was heavily invested given the effort to develop these elite troops.[46] There was an imperative to act, combined with another pressure: time.

Scarcity

Browning, given the situation that existed in September 1944, can be seen as experiencing the same scarcity pressure as Montgomery; he was operating within the same sense of victory euphoria, the overly optimistic sense that the war would be over soon. The main problem arising from this shared sense of optimism meant the plan for Market Garden was moulded to fit a flawed 'the war is likely to end soon' premise.[47] There was also another, more specific aspect to the victory euphoria, with the war seemingly coming to a close, the window of opportunity to use FAAA in the strategic role for which it was intended was closing. Time to return to Browning's Headquarters.

Browning's Headquarters, Moor Park, England, 16th September 1944

Browning was very cognisant of the need to get FAAA into action in a strategic role.[48] Autumn, then winter, would soon set in, making airborne operations highly unlikely if not impossible; if a large-scale operation didn't happen soon, it wouldn't happen at all. The window of opportunity was closing; this window was also closing for him on a more personal level.

Browning having been intimately involved in the raising of the British airborne effort was desperate to lead his troops into battle. He had not held any kind of operational command in the war, so needed one to make career

---

44 Lewis H. Brereton, *The Brereton Diaries – The War in the Air in the Pacific, Middle East and Europe 3 October 1941 – 8 May 1945*, (William Morrow: New York, 1946), 308-309.
45 Clark, *Arnhem*, 97.
46 Powell, *The Devil's Birthday*, 11.
47 Bennett, *A Magnificent Disaster*, xvi.
48 Mead, *General 'Boy'*, 111.

progress after the conflict.⁴⁹ Market Garden represented his last chance of getting into action and gaining an operational command, he needed the operation to go ahead.⁵⁰ The desire to do something about this would lead to rash decisions.

Conclusion

Browning was under a lot of situational pressures to undertake Operation Market Garden. In terms of obedience to authority, there was clear explicit authority, he had been given a lawful order. The underlying politics in FAAA and poor relationships with his American colleagues also made him vulnerable to implicit authority pressures; he would have found it difficult to swim against the tide even if he had wanted to, which he did not. His desire to see airborne forces used in a strategic war winning role and need for an operational command meant he was predisposed to go along with the operation (Commitment and Consistency). This was compounded by the closing window of opportunity as the war looked to be ending, and winter would soon be setting in (Scarcity principle). Browning faced significant external pressures that played on some of his internal drivers, his motives. Browning's motivation for supporting Market Garden is something we will explore in more depth in the next chapter.

---

49 Middlebrook, *Arnhem 1944*, 11.
50 Buckingham, *Arnhem 1944*, 76.

7

Browning's Orientation – Ambition

'Vain, aloof and ambitious'.[1]

Introduction

Browning's Headquarters, Moor Park, England, 16th September 1944

Tomorrow would be a big day for Browning, Market Garden, the biggest airborne operation ever attempted would take place, and he would be in command. It was a vindication of three years of hard work building up the Airborne Forces; it was also, if he allowed himself to admit it, his chance to confirm his reputation and gain an influential position in the post war British Army. Some of his contemporaries saw him as something of an 'empire builder'; if this was the case, this was his chance to achieve his goals.

Goals

Just what were Browning's personal goals for Market Garden? To answer this question, this chapter will examine the personal filters he used to make sense of the situation at the Orientation stage of the OODA Loop; it focuses on a framework of motivational drivers developed by David McClelland. The role that these drivers play in managing difficult situations has been extensively researched. So, let's look at them in more detail.

Motivation

Motivation is seen as shaping your behaviour by first energising you to act and then directing your behaviour towards a particular goal to satisfy the

---
1 Clark, *Arnhem*, 67.

energised need. Your goal is determined by the need that is most relevant or salient to you at the time.[2] Within the OODA Loop model, your motivational state will affect how you orient to the situation, as your sense making will be shaped by which need is most important to you. Of relevance to our discussion, research has examined the motivational drivers of political leaders when engaged in crisis situations that had the potential to lead to conflict or that did in fact result in war.[3] These studies have focused on three basic motivational drivers: the need for Achievement (referred to as the Achievement motive); the need for Affiliation (the Affiliation motive); and the need for Power (Power motive).

Development

The three motive systems we are interested in here were derived from Henry Murray's original work on motivation back in the 1930s, and have since been further refined by a group led by McClelland.[4] The motives have traditionally been measured by the content analysis of imaginative thought using the Thematic Apperception Test (TAT) which presents respondents with a set of pictures about which they write short stories detailing their interpretation. This is what is known in psychology as a projective measure as you 'project' your underlying characteristics, needs or views through the interpretation of an ambiguous stimuli. If you undertake the TAT, you are required to describe what is happening in the picture; your stories are then coded for imagery related to the various needs in question. An example of a picture used in the test would be two scientists in a laboratory, seemingly in conversation.

In this example, if you are strongly motivated by a need for Achievement, you might describe this scene as two scientists who have just found a cure for some kind of disease; the theme is one of success or achievement. Alternatively, you might describe the picture in terms of a supervisor rebuking a subordinate for making a mistake; here the theme would be one of Power (in this case the delineation of a hierarchical relationship). Alternatively, you might tell a story of two friends discussing

---

[2] David McClelland, 'How do self-attributed and implicit motives differ?' in *Motivation and personality – Handbook of thematic content analysis*, ed. Charles Smith, (New York: Cambridge University Press, 1992), 73.
[3] David Winter, 'Content analysis of archival materials, personal documents, and everyday verbal protocols' in *Motivation and personality – Handbook of thematic content analysis*, ed. Charles Smith, (New York: Cambridge University Press, 1992), 110-125.
[4] David McClelland, 'How do self-attributed and implicit motives differ?', 73.

where they are going to socialise that evening; this would be interpreted in terms of demonstrating a need for Affiliation. The story you describe is taken as representative of your underlying or implicit motives. Let's look at each of these three motives in turn.

Need for Achievement

The Achievement motive is defined as a concern with doing things better and surpassing standards of excellence. If you have a high need for Achievement, you will demonstrate a strong need to accomplish goals and to perform to a high standard, you will tend to focus on the task at hand and what needs to be done to achieve a functional outcome. Research studies have shown that if you have a strong Achievement motive, you will work best when the task is moderately difficult. You will try hard to improve on past performance and take moderate risks to get ahead. You will also persist in attempting to reach goals and challenge yourself and others to do better. You will perform better at tasks if you see these as leading to the attainment of your future goals and worse when the outcome is due to chance, or you have no personal responsibility for the task. When prompted, you will recall more uncompleted rather than completed tasks and generally prefer moderately difficult tasks to those that are too easy (and therefore provide no satisfaction upon completion) or too difficult (and are thus not achievable).[5] On the other hand, it also means you can be blinkered in your outlook at times. The Achievement motive then, as the name suggests, is about getting things done.

Need for Affiliation

The Affiliation motive is defined as a concern for establishing, maintaining, or restoring a positive affective relationship with another person or group of persons. If you are high on the Affiliation motive, you will generally form close attachments with other people and consider the needs of those around you. You are likely, therefore, to value friendships, pay attention to the feelings of others and be generally sympathetic towards other peoples' problems. You also tend be consultative in your approach and avoid getting into arguments. Conversely, you can also be accommodating and submissive and may value

---

[5] David McClelland & Richard Koestner, 'The Achievement Motive' in *Motivation and personality – Handbook of thematic content analysis*, ed. Charles Smith, (New York: Cambridge University Press, 1992), 143-152.

goodwill more than logic or reason. You can also be procrastinating. A strong Affiliation motive means you prefer to work with people that you know, like and trust (but are not necessarily competent) rather than unknown or disliked outsiders.[6] With the Affiliation motive, the focus is on relationships, not the task.

Need for Power

The Power motive is defined as a concern with establishing, maintaining and enhancing control of the means of influencing others. If you have a high need for Power you will tend to focus more on your own prestige, valuing your reputation with others. You will also demonstrate a need to convince another person in an argument that you are correct, winning the argument, possibly at the expense of the relationship (Affiliation) or getting the job done (Achievement). You will also prefer (to be at the top of) dominant-submissive relationships. You are more likely to be hierarchical in outlook and be prone to being dominant and argumentative. You will also be more active politically and status-conscious, with a strong concern about your personal reputation and prestige. You will also be exploitative and blaming of others for failure and less willing to compromise in an argument.[7] Controlling matters and your status are thus the key issues.

In summary, how you make sense of a situation will depend on what motivational driver is most important to you: the task (Achievement); people around you (Affiliation); or you (Power). This triumvirate broadly fits John Adair's Action Centred Leadership model, which focuses on Task, Team or Individual needs.[8] The next question is how do these motives affect military and political decision-making?

Research Application

Several studies have explored the role of the three different motives in the management of international crises and military conflicts. For example, one study examined the foreign policy stances of various heads of state.

---

[6] Richard Koestner & David McClelland, 'The affiliation motive' in *Motivation and personality – Handbook of thematic content analysis*, ed. Charles Smith, (New York: Cambridge University Press, 1992), 205-210.

[7] Joseph Veroff, 'Power motivation' in *Motivation and personality – Handbook of thematic content analysis*, ed. Charles Smith, (New York: Cambridge University Press, 1992), 278-285.

[8] John Adair, *Action-Centred Leadership* (London, Gower, 1979).

Transcripts of press conferences given by 45 world leaders were scored for Affiliation and Power motivation. Power motivated leaders were found to be more independent and confrontational in their foreign policy stance, whereas Affiliation motivated individuals were found to promote a more co-operative and interdependent foreign policy.[9]

Another study examined how these motives related to the management and outcomes of international crises; it involved scoring three sets of historical materials, the British Sovereign's Speech at the Opening of Parliament, inter-governmental communications between Britain and Germany immediately before the outbreak of the First World War, and the statements and letters passed between Kennedy and Khrushchev during the Cuban Missile Crisis. The study found that combined Power minus Affiliation motive scores were significantly higher in the years before Britain entered a war in comparison to years of peace. This combined score also increased rapidly during the crisis leading up to the First World War but decreased from the beginning to the end of the Cuban Missile Crisis. British wars tended to end only after Power motive levels dropped.[10]

Another study scored the inaugural addresses of different American presidents from George Washington to Ronald Reagan. These speeches were assessed for themes of Achievement, Power and Affiliation motivation using a content analytical scoring technique. The results indicated that those presidents who were (independently) rated as 'great' scored high on the Power motive, however, they were more likely to involve the United States in a war. Conversely, those presidents who scored highly on the Affiliation motive tended to focus on peaceful resolutions to crises but were also vulnerable to self-serving subordinates.[11] The results of this study also suggested that those presidents who scored highly on the Achievement motive were led by frustration of their initiatives into political intrigue.

A further study scored President Woodrow Wilson's speeches made during three key decision-making events, the ratification of the Versailles Treaty, the US intervention in the Mexican Revolution, and the entry of the United States into the First World War. Wilson's judgements in relation to

---

[9] Margaret Herman, 'Explaining foreign policy behaviour using personal characteristics of political leaders', *International Studies Quarterly*, 24, (1980), 7-46.
[10] Winter, 'Content analysis of archival materials, personal documents, and everyday verbal protocols', 110-125.
[11] David Winter, 'Leader appeal, leader performance, and motive profiles of leaders and followers: A study of American presidents and elections', *Journal of Personality and Social Psychology*, 52, (1987), 196-202.

these three issues were independently assessed by political scientists as examples of poor decision-making. The focus of this study was the relative frequency of the Achievement, Power and Affiliation motives in the speeches that he gave on the relevant issues. In all three situations, Wilson scored highly on the Power motive with only moderate need for Achievement and a low score for Affiliation.[12]

The different studies discussed above are only a snapshot of the research in this area but serve to illustrate that underlying motives impact on how we make sense of a situation. Depending on motivation, we may interpret the same situation as being most relevant or threatening to either the attainment of a particular functional goal (Achievement), relationship with another person or group (Affiliation), or personal reputation/authority (Power). This assessment will then affect actions in relation to dealing with the situation. An understanding of the key motivational drivers that were most important to Browning will therefore provide a useful framework through which to examine how he oriented to the situation and how this impacted his subsequent actions and decisions.

Browning's Motivation

The central issue that we seek to explore about Browning in this part of the book, is the overarching impression of a highly ambitious manipulator who pushed through Market Garden to serve his own purposes. We have already seen glimpses of Browning's ambition in the previous chapter; this suggests his predominant motivation was the need for Power. Let's examine the extent to which this is the case using the framework of the Power motive.

Browning's Need for Power

The Power motive is defined as a concern with establishing, maintaining and enhancing the means of influencing others. Associated with this consideration for control and influence is a concern for enhancing prestige and status. People who have a strong Power motive are also highly ambitious; they can also be manipulative and politically active. Are these traits visible in Browning?

---

[12] Stephen Walker, 'Psychodynamic processes and framing effects in foreign policy decision making: Woodrow Wilson's operational code', *Political Psychology*, 16, 4, (1995), 697-717.

Browning's Disciplinarianism

Browning seems to have been a strict disciplinarian. This trait can be seen early during his time at Sandhurst where, as one cadet commented that when returning late, 'None of us were too anxious to be caught by our Cadet Under-Officer' Browning.[13] Browning continued to emphasise discipline after entering the Army as a Grenadier Guards officer, becoming known to his subordinates as a 'strict disciplinarian with an occasionally explosive temper' whilst Adjutant of 1st Battalion,[14] and again later in his Guards career.[15] He again showed this tendency when he returned to Sandhurst as Adjutant between 1924 and 1928, becoming known as a stickler for discipline and demanding absolute adherence to orders.[16] Later, at the first meeting with his senior officers when Browning was appointed as a Brigadier to command 128th Brigade, two of his three battalion commanders took their seats when they should have remained standing. Browning summoned his Brigade Major ordering him to take the two offenders out of the office and instruct them in the correct protocol.[17] As his biographer states, this incident, and his insistence on discipline more generally, can be seen in Power motive terms as establishing a dominant-submissive relationship, with Browning in the dominant position. Browning carried this insistence on discipline into his role with the airborne forces. He was not pleased by the relaxed attitude to discipline he found when taking up his position as Commander Para-Troops and Airborne Division in November 1941. He promptly brought in a cadre of NCOs from the Brigade of Guards to address the problem, much to the disgust of the airborne soldiers. The Guards are, of course, known for their discipline and so this might simply be a product of that culture, but Browning does appear to be a good example of this. Because of his disciplinarian streak, he certainly had the power of command, something demonstrated early in his career as a young subaltern in the trenches during the First World War.[18]

---

13 Mead, *General 'Boy'*, 12.
14 Ibid., 30.
15 Ibid., 42.
16 Ibid., 36.
17 Ibid., 58.
18 Mead, *General 'Boy'*, 29.

Browning's Control

Browning had, by 1944, established overall control of the British airborne troops; his position in the chain of command left his divisional commanders with little input into and influence over operational planning. This was evident in the planning for D-Day where Browning often represented British airborne interests on his own. More crucially, this carried over into Market Garden where at the first FAAA planning conference, Browning again was the only British airborne representative present. Ritchie suggests this approach was due to his 'determination to emphasise his seniority and absolute authority within the British airborne hierarchy' along with 'his personal responsibility for directing British airborne operation plans'. Everything had to be channelled through him, meaning if Urquhart had harboured any concerns about the operation (which, as we shall see, he did), he was effectively prevented from approaching other senior officers such as Brereton, Dempsey or even Montgomery.[19] Browning's emphasis on discipline and power of command along with his controlling practices suggest he was, in terms of a concern for controlling and influencing others, strongly Power motivated.

Browning's Prestige

The Power motive also refers to a concern for prestige and status. Several incidents suggest that this was an issue for Browning. Browning's first directive, when he took up his new appointment at HQ Airborne Troops on 4th December 1941, was to stipulate that his new command should be referred to as the 'Airborne Corps and I will be referred to as the Corps Commander'.[20] Later, when Brian Urquhart, his Corps Intelligence Officer, applied for a posting away from him, Browning requested he keep their disagreements confidential (to protect his reputation). Buckingham goes as far to suggest Browning saw Market Garden as a chance to enhance his own reputation and cement his place in the post-war British Army.[21] Browning was also, but perhaps of lesser significance, very particular about his appearance; even whilst in the trenches during the First World War he was seen as an

---

19 Ritchie, *Arnhem*, 180-181.
20 Mead, *General 'Boy'*, 96.
21 Buckingham, *Arnhem 1944*, 77.

'exceptionally well turned out officer',[22] firming up this reputation when he mounted a guard of honour which was excellent even by Guards standards.[23] On a related note, he was 'in his element' when it came to drill, viewing it as an essential part of discipline.[24] Mead, has the final say on this; he states in his biography that by mid-1942 his subject had became known as 'Bullshit Browning' and was 'more concerned with burnishing his own reputation than fighting the Germans'.[25] Prestige, reputation and status, do seem to have been important to him, which leads us to the fourth aspect of the Power motive, politicking and manipulation.

Browning's Manipulation

Browning was a sharp political operator.[26] An old Etonian, he had good connections which he used to his advantage, his father certainly used his own connections to get Browning into the Grenadier Guards.[27] Browning, incidentally, had served briefly with Winston Churchill in the First World War, where he gave the future Prime Minister his greatcoat when showing him around the trenches.[28] Most commentators cite Browning as using his political connections and skill to help establish the airborne forces, they also, however, cite this characteristic as playing out in a manipulative streak. This is often evidenced in the removal of Eric Down from command of 1st Airborne Division and posting to India, which, like most things related to Market Garden is quite contested.

Mead suggests Browning was not responsible for Down's removal and indeed contested it,[29] but other commentators disagree. Buckingham argues that there was no pressing need to send Down to India, indeed after his removal he spent time 'kicking his heels' in a staff appointment in Britain. Buckingham goes on to suggest Browning did not like Down who was a vocal critic with more operational experience than him and so had him removed. Gerald Lathbury, a brigadier within the division had been promised the position but Buckingham argues that he was too experienced

---

22 Mead, *General 'Boy'*, 13.
23 Ibid., 25.
24 Ibid., 13.
25 Ibid., 76.
26 Clark, *Arnhem*, 68.
27 Mead, *General 'Boy'*, 10-12.
28 Ibid., 16.
29 Ibid., 96.

to be acceptable to Browning. Roy Urquhart's subsequent appointment was seen as a surprise due to his lack of airborne experience. He was, however, a Montgomery protégé, (Montgomery knew and trusted him), and the British Army did operate a one size fits all approach to command. Buckingham concedes Browning may have accepted Urquhart to please Montgomery but argues Urquhart would also be beholden to Browning and his lack of airborne experience meant that Browning could control him. There is certainly a fair chance that Down is unlikely to have accepted the plan for Market Garden imposed by Browning. Finally, Buckingham concludes that the appointment of a 'tame outsider' would send a message to the division as to where the 'power lay'.[30]

Browning's Meddling

Buckingham also suggests that Browning was a meddler as well as a manipulator. He cites the Bruneval Raid in February 1942 and his imposition of a rigid and inflexible plan on Major John Frost, the officer commanding the operation. When Frost objected, he was told he would be removed if he did not comply (Frost's concerns later proved justified). Having dictated the plan, Browning then left his liaison officer to deal with Frost, a move Buckingham interprets as a step to distance himself in case of failure of the operation. Buckingham argues that he repeated this behaviour in September 1944, on both occasions he 'oversaw the imposition of unsuitable plans with no regard for the opinions of those tasked to carry them out' whilst he also 'distanced himself from possible repercussions'. Buckingham is clearly no fan of Browning, summing him up as a ruthless manipulative empire builder.[31] If we follow Buckingham's line of argument (it is hard not to), Browning certainly comes across as a politically adept manipulator. This just leaves the final aspect of the Power motive, ambition.

Browning's Ambition

Mead agrees Browning was personally ambitious but genuinely wanted the airborne concept to succeed.[32] Buckingham, on the other hand, portrays him as an opportunist who saw in the airborne forces a chance to further his

---
30 Buckingham, *Arnhem 1944*, 26-31.
31 Ibid., 16-17.
32 Mead, *General 'Boy'*, 76.

own career. He contests Browning's label as the 'father of airborne forces', pointing out that he was appointed when the capability was expanding not when it was founded. Browning, he argues, with no operational command in the war, was appointed because a political operator was most needed.[33] Ritchie suggests Browning agitated for expansion of the British airborne forces to create the need for a higher command position, to which he would be appointed.[34] Mead does accept that some of Browning's peers found him pushy; when George Hopkinson was killed in Italy in 1943, Browning, present in the country at the time, asked if he should assume command of 1st Airborne, but was told no.[35] Earlier, in 1942, after examining airborne operations in North Africa, Browning recommended establishing the role of airborne advisor at Allied Force Headquarters; Ritchie argues this was a 'transparent attempt of a very ambitious officer to secure his own appointment to the new position'.[36] In terms of Market Garden, Buckingham suggests Browning chose British 1st Airborne for the 'prize' at Arnhem as this objective had the most kudos.[37] Ritchie concludes that by September 1944, Browning was desperate to get British I Airborne Corps into the action as he needed to justify the two divisions under his command and thus his rank of Lieutenant-General.[38]

Browning's behaviour also gave the impression to his American airborne colleagues that he was 'out for his own advantage'.[39] They certainly distrusted him, believing that he knew all the right people and used them to his own advantage; viewing him as a devious manipulator.[40] Matthew Ridgway was certainly a threat to Browning as he had more airborne experience and operational command experience.[41] Ridgway went so far as to warn James Gavin, when appointed to command US 82nd Airborne in his stead, about Browning's 'machinations and scheming'.[42] This view is supported by Ray Barker, the COSSAC Deputy Chief of Staff, who also warned Gavin that Browning was an 'empire builder'.[43] Browning seems

---

33 Buckingham, *Arnhem 1944*, 15-16.
34 Ritchie, *Arnhem*, 98.
35 Mead, *General 'Boy'*, 94.
36 Otway, *Airborne Forces*, 81.
37 Buckingham, *Arnhem 1944*, 80.
38 Ritchie, *Arnhem*, 100.
39 Mead, *General 'Boy'*, 101.
40 Bennett, *A Magnificent Disaster*, 13.
41 Buckingham, *Arnhem 1944*, 18.
42 Powell, *The Devil's Birthday*, 38.
43 Mead, *General 'Boy'*, 101.

to have been imbued with the strong sense of ambition that is part of the Power motive.

Conclusion

In summary, Browning appears to have been strongly motivated by a need for Power. This would mean that he was controlling, overly concerned about his reputation, manipulative and ambitious. The evidence suggests that he was all of these; Clark has described him as 'vain, aloof and ambitious'.[44] This is not just a judgement made with hindsight, his Americans colleagues viewed him as an empire builder at the time and warned each other of this. His professional involvement in the development of airborne forces meant that he was keen to see them come to fruition. Lacking an operational command in the war, he was even more desperate to get into the action before it was too late. His professional and personal ambitions would have affected how he appraised the situation he faced by shaping his personal goals at the time. The challenge to these goals would, in turn, have shaped how he assessed the risks associated with Market Garden and thus, in terms of the Decision Conflict Model, the coping strategy he would adopt in the decision-making phase of the OODA Loop. We will examine this in more detail in the next chapter.

---

[44] Clark, *Arnhem*, 67.

8

Browning's Decision – Bolstering

'The Red Devils and the Polish Paratroopers can do anything'.[1]

Introduction

Browning's Headquarters, Moor Park, England, 10th September 1944

Major-Generals Roy Urquhart and Stanislaw Sosabowski were sat in their commander's office near Rickmansworth in England. Having worked through the planning for Operation Comet, Sosabowski had raised with Urquhart his deep reservations about the enormity of the task allotted to his brigade. Urquhart had listened to his concerns and then suggested they present them to Browning. Browning listened to Sosabowski's concerns, he agreed, but added that he had no more troops available. When Sosabowski pressed his case again, Browning closed the conversation down with the rather glib statement 'The Red Devils and the Polish Paratroopers can do anything'.[2] Browning, under pressure to go ahead with the operation, was appearing to make light of the matter; why was this?

Coping strategy

In this chapter, we will discuss the coping strategy that Browning adopted in the Decide stage of the OODA Loop, utilising by way of comparison with Montgomery, the Decisional Conflict Model. It will be useful, before we start working our way through the model, to consider the background to the situation that Browning was facing in early September 1944.

---

1 Sosabowski, *Freely I Served*, 143.
2 Ibid., 143.

Browning's Headquarters, Moor Park, England, 16th September 1944

First Allied Airborne Army was and had always been a particularly unhappy place for Browning. Problems had begun at the time that FAAA was formed in August. Much to his annoyance, command of the formation had gone to the US airman, Lewis Brereton, Browning was made deputy; he was not happy as he was slightly more senior and highly familiar with the airborne concept, Brereton was not. As explained earlier, Browning had gotten off to a bad start with his fellow Corps Commander, Ridgway and the situation was not helped by the fact that Ridgway was, essentially, Browning's nearest rival. Ridgway was highly suspicious of Browning, indeed the whole of the American fraternity generally disliked or mistrusted him;[3] they viewed him as arrogant and patronising.[4] Not only was he on thin ice it had also begun to crack with his earlier threat to resign.

Weakened position

Most commentators agree that Browning had severely weakened his position with his threatened resignation and was thus hamstrung in his ability to challenge Market Garden.[5] This would clearly impact on how he coped with the situation he faced and how he assessed the antecedent conditions in the Decisional Conflict Model. The first of these is the perception of risk, so let's examine that now.

Risk

To explore Browning's perception of the risk posed within Operation Market Garden, we need to delve into the murky waters of the intelligence picture and what was known regarding the presence of German armour in the Arnhem area. As we shall see, this is a hotly contested issue, especially Browning's role in (mis)handling the intelligence. Buckingham asserts that Browning 'deliberately suppressed intelligence showing the presence of SS troops and armour near Arnhem'.[6] Harvey goes as far as to accuse Browning of being responsible for omitting details from the Market Garden

---

3   Mead, *General 'Boy'*, 108-109.
4   Powell, *The Devil's Birthday*, 38.
5   Harclerode, *Arnhem*, 49.
6   Buckingham, *Arnhem 1944*, 95.

Intelligence Summary of the 'Panzer troops' previously referenced in the Comet plan. Notwithstanding conspiracy theories of deliberate manipulation by Browning, there is good evidence to suggest that the intelligence reports coming out of Browning's Headquarters 'glossed over any mention of the significant threat posed by II SS Panzer Corps'.[7] The Intelligence Summary for 13th September is quoted as evidence of this with its assertion that the 'troops manning [the defences] are of low category'.[8] The key question, to help unpick this controversy and answer our question about Browning's risk perception, is what did he know about the likely German resistance in the Market Garden area of operations?

Browning had been made aware of II SS Panzer Corps' presence in the operational area on 10th September but, crucially, it seems was not made fully aware of its combat worthiness; it had been described as battered, very understrength and thus did not pose much of a threat.[9] Browning's Corps Headquarters in turn had passed this view on to 1st Airborne Division in a 13th September Intelligence Summary noting the presence of Panzers in the area.[10]

The battle worthiness of both Panzer divisions is something of a red herring as the British intelligence staffs failed to factor in German capacity to generate ad hoc battle groups from lower-grade troops at short notice.[11] These groups would ultimately play key roles in initial German defence and subsequent counter attacks. The point of the controversy, however, is about the German armour, so let's return to this. The presence of the Panzers was also brought to Browning's attention by another source, Photo Reconnaissance (PR).[12] This is where we dive even deeper into the murky world of the Arnhem intelligence debate, and the saga of Brian Urquhart (no relation to Roy), the Chief Intelligence Officer for I Airborne Corps.

Brian Urquhart's Tale

The safe in Major Brian Urquhart's office contained a letter he had written to the Prime Minister laying out his concerns about Market Garden.[13] He had written it because he was a deeply worried man. He was concerned

---

7 Harvey, *Arnhem*, 34.
8 Clark, *Arnhem*, 110.
9 Ibid., 109.
10 Urquhart, *Arnhem*, 9.
11 Kershaw, *It Never Snows in September*, 108-112.
12 Bennett, *A Magnificent Disaster*, 199.
13 Middlebrook, *Arnhem 1944*, 66.

that the desire to get into action was clouding his colleagues' judgements. He did not share the opinion that it would be a walkover. He felt talk of an 'airborne carpet' and the mission as being a 'party' was helping lull people into thinking the operation would be easy. He had become alarmed about the presence of German armour through his reading of the Twenty-First Army Group and Second Army Intelligence summaries, along with accounts from prisoner interrogation and reports from the Dutch resistance.[14] He had ordered a Photo Reconnaissance mission because of these concerns.[15]

It seems Urquhart had ordered the PR mission on 12th September and received the photographs, showing tanks in the Arnhem area three days later.[16] He had raised his concerns to both Browning and Gordon Walch, his Chief of Staff, who were not impressed.[17] Browning dismissed the tanks as 'probably unserviceable'.[18] This is where it gets murky. Urquhart stated after the war that he was subsequently visited by Browning's Chief Medical Officer, Colonel Austin Eagger, who reluctantly sent him on sick leave because he was suffering from 'nervous strain and exhaustion'. This is the generally accepted story, which if true portrays Browning deliberately suppressing unwelcome intelligence. There is, however, some doubt over this version of events.[19] Urquhart may also have been misrepresented by Cornelius Ryan in his book where he recounts the incident with the Medical Officer.[20]

There is also uncertainty over the dates for the Photo Reconnaissance mission. Weather conditions precluded any flights on 14th or 15th September, and Tony Hibbert, a 1st Airborne Division officer claims he saw the photographs on the 12th. This suggests the mission was flown before 10th September, during the planning for Comet. Some doubt even exists as to whether the photographs existed; no record of a low-level PR mission has been found. Also, the low-level imagery would have been extremely lucky to have picked up any tanks and German records show that armour was not actually in the area at the time.[21] That seems to settle that question.

The Arnhem intelligence debate is, however, murky waters indeed. Ritchie argues that the photographs existed but came from high-altitude

---

14 Bennett, *A Magnificent Disaster*, 58.
15 Hibbert, *The Battle of Arnhem*, 38.
16 Powell, *The Devil's Birthday*, 44.
17 Bennett, *A Magnificent Disaster*, 58.
18 Middlebrook, *Arnhem 1944*, 65.
19 Ibid., 66.
20 Buckley & Preston-Hough, *Operation Market Garden*, xvii.
21 Ritchie, *Arnhem*, 133.

Spitfire flights on 12th September and that these were shown to Browning.[22] If that is true, then Browning was aware of some threat from German armour. This would be further corroborated if there was evidence that this intelligence was passed downwards to 1st Airborne Division. This, at first glance does not seem to be the case.

Roy Urquhart states in his memoirs that there was a lack of intelligence to support Market Garden, that his staff at the time were 'scratching around for morsels'.[23] Frost confirms at divisional level they were told German resistance would be light, composed mainly of under-equipped SS recruits and Luftwaffe personnel.[24] Lathbury would prepare the plan for 1st Parachute Brigade on the assumption of minor enemy opposition.[25] We are, however, in murky waters as it appears, some details were passed on.

Roy Urquhart, (we shall continue to use his full name here to avoid confusion with Brian), states that he was under no illusions that the Germans would 'fold up at the first blow'.[26] He states that Browning himself told him about the German formation in the area being a Brigade group plus a few tanks, briefing this at his Orders Group on 12th September.[27] Roy Urquhart also states that his brigadiers were told that two depleted SS Panzer divisions were located near Arnhem.[28] Charles Mackenzie, his Chief of Staff would later recall vague stories of German tanks refitting in the area.[29]

The likely presence of German armour near Arnhem does appear in intelligence reporting. A 1st Airborne Division Intelligence Summary, dated 7th September, had previously mentioned 50 tanks and a 'fair quota of Germans' in the area;[30] a later divisional report (14th September) noted tanks that had been 'previously reported'.[31] Further down the chain of command, a 1st Parachute Brigade Intelligence Summary, dated 13th September noted that the Arnhem area was being prepared for defence and referenced two

---

22 Ibid., 134.
23 Urquhart, *Arnhem*, 7.
24 Frost, *A Drop Too Many*, 200.
25 Ritchie, *Arnhem*, 141.
26 Urquhart, *Arnhem*, 9.
27 Ritchie, *Arnhem*, 136.
28 Bennett, *A Magnificent Disaster*, 198.
29 Ibid., 59.
30 TNA, WO 171/341, 1st Airborne Division War Dairy, Planning Intelligence Summary 2 of 7, September 1944.
31 TNA, AIR 37/1217, Operation Market, 1st Airborne Division Planning Intelligence Summary, No.2 dated 14th September 1944, prepared by G2(I), 1st Airborne Division, 14 September 1944.

'battle-scarred Panzer divisions reforming in the area'.[32] Tony Hibbert also states he and other Brigade Majors in the division were briefed on the presence of II SS Panzer Corps in the area and shown photographs of 'German panzer IVs ... tucked in underneath woods' (presumably the reconnaissance photographs ordered by Brian Urquhart).[33] Furthermore, Buckley points out that allied commanders, from battalion level upwards, had been briefed about the possibility of SS units in the area.[34]

Contrary to the 'myth' of withheld intelligence, 1st Airborne Division does seem to have had a broadly correct picture of the German threat at Arnhem; that there were large formations, with tanks, within 20-30 miles of Arnhem and that this included 9th SS Panzer Division.[35] It seems enough intelligence had been received to cause disquiet within the division; Roy Urquhart certainly expected the Germans to put up resistance.[36] Sosabowski also claims that he expected bitter fighting.[37] Lathbury and Hackett briefed their commanders to expect 50 per cent casualties,[38] the latter telling his battalion commanders that the hardest fighting would be involved in getting to the bridges, not holding the perimeter.[39] Indeed, Hackett later claims when asked when he first knew that the operation would be a disaster, his response had been 'before the battle started'.[40]

It seems then, that as Powell argues 'there can be no grounds for any suggestion that [Browning] deliberately hid information, either from General Urquhart or from anyone else'. As Ritchie points out, there is evidence that Browning did paint a 'discouraging picture' of likely German resistance; Lathbury confirms this, stating that the presence of Panzers in the Arnhem area was mentioned by Browning during a Comet briefing.[41] It should be remembered that Browning had worked to enlarge the previous single division Operation Comet, to the three and a half division Market Garden, due to perceptions of increased threat. This is where we turn to another

---

32 TNA, WO 219/5137, 1 Parachute Brigade Intelligence Summary No.1, by Capt W.A. Taylor, 13 September 1944.
33 Ritchie, *Arnhem*, 134.
34 Buckley & Preston-Hough, *Operation Market Garden*, 216.
35 Ritchie, *Arnhem*, 141.
36 Harclerode, *Arnhem*, 54.
37 Sosabowski, *Freely I Served*, 146.
38 Buckingham, *Arnhem 1944*, 93.
39 Baynes, *Urquhart of Arnhem*, 100.
40 Harvey, *Arnhem*, 51.
41 Ritchie, *Arnhem*, 136.

debated point and perhaps the most famous quote about Market Garden, so famous they named a book and film after it.

A Bridge Too Far

Part of the folklore surrounding Arnhem is the comment about going 'a bridge too far' that is generally attributed to Browning upon being briefed by Montgomery on Market Garden on 10th September. The narrative around this event suggests Montgomery told Browning he would need to hold Arnhem bridge for two days, to which Browning replied: 'We can hold it for four, but I think we might be going a bridge too far'.[42] This story first outlined in Roy Urquhart's 1958 memoirs has been debated since. Gordon Walch, Browning's Chief of Staff, who saw him straight after the meeting, certainly believes he said it, pointing out that he also qualified the statement with the rider 'with the airlift available'.[43] It seems, then, that Browning was sufficiently aware of the risks posed by German opposition in the Arnhem area as to hold reservations about Market Garden.

Browning's estimate of the extent of the likely German opposition does, however, need to be considered against the background context. We have examined the phenomenon of the victory euphoria that existed within the Allies at the time; Browning was as guilty of this as anyone else. The highly experienced Sosabowski certainly argues Browning appeared overconfident and clearly underestimated the enemy.[44] Ritchie, conversely, points out that if Browning did underplay the issue of II SS Panzer Corps, then he was simply keeping in line with Montgomery and other senior officers at Twenty-First Army Group.[45]

In summary, it seems that Browning was aware of the threat posed by the likely German opposition to the operation. He was told about II SS Panzer Corps on 10th September, although its fighting ability was underplayed.[46] Bringing discussion back to the Decisional Conflict Model and Browning's likely response to the first question it poses about risk, this suggests Browning was aware of the threat posed by the likely German opposition to the operation. We would argue, therefore, that he did assess there was risk

---

42 Powell, *The Devil's Birthday*, 32.
43 Mead, *General 'Boy'*, 115-116.
44 Sosabowski, *Freely I Served*, 142.
45 Ritchie, *Arnhem*, 134.
46 Clark, *Arnhem*, 109.

in the situation. The next question to consider, in DCM terms, was whether he believed that the operation was feasible.

Feasibility

Browning, as an airborne officer was clearly heavily involved in the air component (Market), and so we will now focus our discussion here on this side of the operation. Browning's sense of the viability of Market would have been largely dependent on the plan for the air lift as this dictated the extent of the operation and how it would be undertaken. Many commentators point to the air plan as one of the main reasons for the failure of the operation and as the air lift was under the control of the air forces, blame Allied air planners. This is where we delve back into another controversy and, some would argue another myth that has been generated about Market Garden.

Air Lift

It is true that the air lift was under the control of the air forces. For the British, this had been enshrined in a War Office memorandum that stated in 1943 that airborne operations were an air operation until the troops landed on the ground.[47] Ritchie argues that this was a sensible arrangement as the air commander would be best placed to oversee the plan for the air lift and undertake the necessary coordination to ensure it happened.[48] This arrangement has meant that the air force planners have attracted a great deal of criticism for the plan they put together and are indeed a frequent target for the blame for the failure of the operation.

Roy Urquhart, in his After-Action Report points to the multiple lifts across three days along with the selection of the landing and drop zones (and lack of Close Air Support) as key causes of the difficulties at Arnhem.[49] His criticisms are echoed by several authors who have contributed to the wide body of historical literature on the subject. Buckingham, for example, argues that the RAF Planners were too focussed on their own needs, overly concerned about losing aircraft.[50] Roger Cirillo agrees, arguing senior airmen 'safe sided' their part of the operation.[51] Harclerode focuses on the selection

---

[47] Buckingham, *Arnhem 1944*, 9.
[48] Ritchie, *Arnhem*, 53.
[49] AHB, 1st Airborne Division Report on Operation Market, 10 January 1945.
[50] Buckingham, *Arnhem 1944*, 87.
[51] Cirillo, *Market Garden and the Strategy of the Northwest Europe Campaign*, 56.

of the landing and drop zones without airborne involvement;[52] we will, however, explore this issue in more detail in the next part of the book. Ritchie, conversely (perhaps unsurprisingly as the air force historian), challenges these criticisms, referring to the RAF as scapegoats;[53] indeed, he goes so far as to refer to the air lift issue as one of the myths about Market Garden.[54] He points out there is no record of British airborne commanders raising objections about the plan. He also argues, very persuasively, that given the circumstances, the air lift plan was correct and 'in no sense placed the needs of the air forces above those of the men being delivered into battle'.[55] So, what were these circumstances?

The first point to note is the short timescale within which the air planners had to operate, which meant the planning was rushed. That this was the case was evident to some in the US 101st Airborne Division who had a sense that the plan was put together in great haste.[56] A key impact of the short timescale was Brereton's decision early in the planning process, that major decisions already taken would need to stand and could not be modified.[57] This would mean that any challenges to the plan were rejected on this basis. Fortunately, or perhaps unfortunately, the planners were helped by the existence of plans from previous cancelled operations.

The air plan for Market Garden was based 'in principle' on the aborted Operation Linnet; Ritchie argues that Browning played a key role in this decision and that it was only possible to mount Market Garden in the required timeframe by using the Linnet plan.[58] The problem was that this plan called for double towing of gliders over a shorter range to ensure a double lift on the first day.[59] This would not be possible because of the greater transit distance to the Market Garden area. Brereton had warned about the problems that would be involved with the greater distance to Arnhem, but his warnings were ignored by Montgomery who went straight to Eisenhower to get approval.[60] These problems were compounded by addition of the two American airborne divisions, which meant that there was simply not enough

---

52 Harclerode, *Arnhem*, 161.
53 Ritchie, *Arnhem*, 252.
54 Ibid., 19.
55 Ritchie, *Arnhem*, 203-204.
56 Bennett, *A Magnificent Disaster*, 33-36.
57 Ryan, *A Bridge Too Far*, 103.
58 Ritchie, *Arnhem*, 251.
59 TNA, WO 219/4998, Operation Sixteen Outline Plan, 10 September 1944.
60 Hamilton, *The Battles of Field Marshal Bernard Montgomery*, 451.

available transport aircraft to meet the requirements.⁶¹ Linnet and Comet, the immediate predecessor of Market Garden, required much fewer aircraft. Added to all of this was the shorter daylight hours available in mid-September that would restrict operations.⁶² This not only restricted flying hours but meant less time to turn the aircraft around and undertake maintenance or repairs; this problem was compounded by a lack of groundcrew, this supporting arm had not been expanded in line with the increase in aircraft and aircrew.⁶³ Added to this challenge, US aircrews were not properly trained for night flying and so would be unable to operate early in the morning or into the evening.⁶⁴ There were also other considerations. The middle of September was a no moon period which essentially ruled out a night drop.⁶⁵ The scattered drops during the D-Day operation suggested a daytime lift would achieve the desired concentration of force;⁶⁶ indeed, the Arnhem drop in particular would achieve a very good concentration.⁶⁷ Finally, operating during the day also meant the Allies would have air superiority and thus be able to protect the vulnerable transport aircraft from fighter interception.⁶⁸ Paul Williams, the head of the US IX Troop Carrier Command, taking these factors into consideration, (probably correctly), decided on only one air lift per day.⁶⁹ He was supported in this decision by Brereton.⁷⁰ This would, however, have serious implications for the operation.

The decision to undertake only one air lift a day meant the airborne divisions would have to be inserted over several days, rather than just one. The operation was viewed as a 'bottom to top' undertaking with the initial objectives being given priority. The logic behind this was that there was no point in seizing the bridges at Arnhem (at the top of the map) if the Nijmegen ones (in the middle) were not taken, similarly there was no point seizing these bridges if those at Eindhoven (at the bottom) were lost. This meant that the plan involved putting US 101st Airborne Division down at Eindhoven in one lift on the first day, with US 82nd Airborne Division inserted in two lifts on

---

61 Buckingham, *Arnhem 1944*, 82.
62 John Warren, *Airborne Operations in World War II, European Theater* (USAF Historical Division, Research Studies Institute, Air University, 1956), 92.
63 Warren, *Airborne Operations in World War II*, 224-226.
64 Otway, *Airborne Forces*, 81.
65 Harclerode, *Arnhem*, 48.
66 Ritchie, *Arnhem*, 182.
67 Warren, *Airborne Operations*, 90.
68 Ryan, *A Bridge Too Far*, 108.
69 Bennett, *A Magnificent Disaster*, 34.
70 Mead, *General 'Boy'*, 118.

successive days (first day and second day) and British 1st Airborne Division (and the Poles) coming in over three days (first, second and third).[71] This would cause serious problems for both the American 82nd and especially British 1st Airborne Divisions. Regarding the latter, Middlebrook points to 'the refusal to fly two lifts on the first day, which resulted in the prolonged dispersal of the division and its failure to achieve concentration of effort'.[72] Powell quotes a senior RAF Officer's criticism of the plan: 'The Air Plan was bad. All experience and common sense pointed to landing all 3 Airborne Divisions in the minimum period, so that they could form up and collect themselves before the Germans reacted'.[73] The weakening of the airborne divisions' ability to assemble sufficient combat power on the crucial first day, brought the viability of the entire operation into question. This was made more problematic by the selection of the landing and drop zones, especially at Arnhem, but we will discuss this in more detail later. The decision to operate in daylight led to another problem.

The Comet plan, upon which Market Garden was based, included *coup de main* glider operations very close to the target to seize the road bridge at Arnhem.[74] This style of operation was very much in line with airborne doctrine and practical experience. Major-General Richard Gale, the commander of British 6th Airborne Division, when consulted on the subject, stated a *coup de main* operation was vital to gain the required element of surprise.[75] A daylight drop, however, meant gliders would be too vulnerable to the reported increasing density of anti-aircraft artillery around the target and so this element of the operation was dropped.[76]

In summary, dispassionate analysis, based on hindsight, of the problems with the air plan, namely, the air lift taking place over multiple days, landing and drop zones located a distance from the objectives and no *coup de main* operations, put the feasibility of the operation in doubt. The key question, that will allow us to judge Browning's response to the second question in the Decisional Conflict Model, is how much of these issues did he appreciate at the time?

---

71 Middlebrook, *Arnhem 1944*, 14-16.
72 Ibid., 443.
73 Powell, *The Devil's Birthday*, 235.
74 TNA, WO 219/4998, Operation Sixteen Outline Plan.
75 Mead, *General 'Boy'*, 120.
76 Buckley & Preston-Hough, *Operation Market Garden*, 215.

Browning's Headquarters, Moor Park, England, 17th September 1944

Browning was aware of the problems with Market Garden. Urquhart had 'badgered' him for more aircraft when briefed on the air plan at 0900 on 11th September.[77] Urquhart duly protested but Brereton had backed Williams' decision about one lift per day.[78] Even Montgomery had gotten involved when informed of the decision on 15th September, sending Brigadier Belchem to remonstrate with Brereton and try to get the number of lifts reduced to two, but he was unsuccessful.[79] Cirillo correctly argues that Williams' decision effectively 'restructured the ground battle' and violated the original intent of the ground commanders as they had designed and understood the plan, Dempsey and Browning were those ground commanders. Browning had been involved in putting the plan together and both he and Dempsey 'were clear as to what they believed would work', and this was not it.[80] Browning had been sufficiently concerned himself about the feasibility of the operation that, he had consulted Gale on the matter, who had also argued for more aircraft, insertion closer to the objectives and *coup de main* operations.[81] Furthermore, upon learning of the issues of the location of the landing and drop zones on 5th September, during the planning for Comet both Dempsey and Browning had recommended switching the operational objective from Arnhem to Wesel. Montgomery had not agreed to this, and so he was left with Arnhem as the final objective.[82]

Browning's Reservations

Putting all this together suggests that Browning did seriously question the viability of the operation as scoped. We would argue, therefore, that his response to the second question in the Decisional Conflict Model, whether the operation was feasible, was negative. If this was the case, he would have adopted the coping strategy of Defensive Avoidance; he was aware of the risk but believed there was not a wholly viable solution. As we have seen in our discussion on Montgomery, the coping behaviours associated with this strategy can take one of three forms: Scapegoating, where the problem is

---

77 Powell, *The Devil's Birthday*, 34.
78 Middlebrook, *Arnhem 1944*, 17.
79 Harclerode, *Arnhem*, 50
80 Cirillo, *Market Garden and the Strategy of the Northwest Europe Campaign*, 45-46.
81 Middlebrook, *Arnhem 1944*, 18.
82 Ritchie, *Arnhem*, 250.

placed in someone else's hands; Procrastinating, where no decision is made; or Bolstering, where the person involved knuckles down and makes the best of a bad job. Which one did browning adopt?

Bolstering

Browning's adoption of a particular coping strategy would be driven not only by a professional appreciation of the military factors, but also his personal motivation. As we saw in the previous chapter, an ambitious (Power motivated) Browning was desperate to get into battle.

Browning had not held an operational command during the war, so therefore needed to gain some operational experience before it ended. With the war looking like it was ending, and with autumn/winter approaching, Market Garden was probably his last chance. Brian Urquhart was certainly of this opinion. Browning's eagerness to be seen in command can be seen in his insistence on landing at Nijmegen (alongside US 82nd Airborne) with his Corps Headquarters (using up 38 valuable gliders in the process).[83] He was, therefore, predisposed to see the operation take place, it was the best of a bad job and better than the alternative, cancellation, but how did this manifest itself in Bolstering behaviours?

Browning had confirmed to Montgomery that an augmented 1st Airborne Division could undertake Comet when a more dispassionate analysis would have concluded that it would have struggled to take and hold the bridges.[84] He certainly did not particularly critique the plan for Comet as a whole when it emerged from Montgomery and Twenty-First Army Group.[85] Ritchie argues Browning simply didn't want to tell Montgomery that it couldn't be done.[86]

Further evidence of Browning's Bolstering behaviour can be seen in his acceptance of the air plan. He did query the air plan, but most commentators suggest that he did not contest it enough.[87] This can be seen in the way he responded to Urquhart's protests, reiterating to the latter at the air plan briefing on 11th September that the allocation of aircraft was necessary as the operation was 'bottom to top'.[88] Buckley argues that Browning also did

---

83 Middlebrook, *Arnhem 1944*, 11.
84 Mead, *General 'Boy'*, 112.
85 Ritchie, *Arnhem*, 100.
86 Ibid., 205.
87 Mead, *General 'Boy'*, 119.
88 Powell, *The Devil's Birthday*, 34.

not press hard enough on the lack of a *coup de main* operation.[89] Browning, after being told by Gale that the Arnhem operation needed a *coup de main* operation, a brigade dropped close to the bridge and more aircraft, asked Gale to keep quiet about his reservations, concerns so strong that Gale would later claim that he would have resigned rather than accept the plan as structured.[90] Browning seems to have put a brave face on matters. Gavin was surprised that he did not question Urquhart's plan when he presented it at a later meeting.[91] Browning outwardly displayed an air of optimism and confidence and was reported to have been in high spirits at the initial briefing for the operation.[92]

In summary it seems there is good evidence to suggest that Browning did, in fact, adopt the Defensive Avoidant coping strategy and thereby engage in Bolstering activity, putting a brave face on the situation. Browning would be so effective at this, as Clark argues, his confidence would be one of the reasons why Market Garden went ahead.[93]

Conclusion

Browning appears to have adopted the Defensive Avoidance coping strategy. He was sufficiently aware, from the available intelligence, of the threat of German resistance in the Arnhem area, which clearly posed a risk to the final objective and the operation. He was also sufficiently aware of the implications of the problems in the air plan as to have serious doubts about the feasibility of Market Garden. There was another possible solution to his doubts, however, cancel the operation. But to do so would remove his opportunity to get into action and have a successful operational command before the war ended; Market Garden was his last, best chance to fulfil his personal goals. Browning's ambition, therefore, coloured his judgement and led to the adoption of the Bolstering form of behaviour. In this mental state he would not have engaged in robust, analytical thinking and would instead be prone to cognitive biases, ultimately making errors in judgement. We will examine these errors and the biases that caused them in the next chapter.

---

89 Buckley & Preston-Hough, *Operation Market Garden*, 215.
90 Middlebrook, *Arnhem 1944*, 18.
91 Ryan, *A Bridge Too Far*, 126.
92 Ibid., 109.
93 Clark, *Arnhem*, 112.

9

Browning's Action – Endowment

Father of the Airborne (?)

Introduction

Browning's Headquarters, Nijmegen, Holland, 26th September 1944

It is late morning on 26th September, Browning is sitting alone in his office in his advanced Headquarters near Nijmegen. Operation Market Garden is over. The remnants of Roy Urquhart's 1st Airborne Division have just been evacuated back across the Rhine in an overnight operation, it looks like just over 2,000 have made it back. Two thousand out of just over 10,000. It's a bloody disaster. It won't be long before the postmortem starts and questions are asked. Browning as commander of the operation, will be on the hook to answer some of them.

Browning's errors

We will look at the different issues as Browning conducts his own postmortem. For this, our focus will be on the final Act stage of the OODA Loop. We will focus down from the strategic level to look at his actions around Market Garden itself, and, more specifically, the airborne component. Although the operational concept was Montgomery's, Browning does bear some responsibility for its flaws. Browning was Montgomery's airborne advisor and there is evidence to suggest his involvement in the plan's development.[1]

---

1 Clark, *Arnhem*, 27.

We will, therefore, examine errors in judgement made during the planning for Market Garden and Browning's role in making them.

Planning Errors

Looking dispassionately at Market Garden, it is hard not to conclude that it was essentially a fundamentally flawed concept. Middlebrook points out that it was not unreasonable to assume that XXX Corps would be able to advance quickly against what had been a weak and disorganised German resistance,[2] but many commentators would seem to disagree. As Ritchie points out, the plan simply replicated the flaws already present in Comet; the objectives were too distant, the ground forces would need to advance along a narrow road across numerous water obstacles. The airborne forces were given too many tasks over a too widely dispersed area having been dropped on unsuitable landing and drop zones.[3] He argues that Market Garden was likely to fail because its objectives were a chain of sequential dependencies and because airborne operations to date had been characterised by operational failure to a greater or lesser extent. The short time scale for planning did not help matters. The flawed concept was further weakened by poor planning,[4] which has been identified by those involved in the operation. Brian Urquhart, as we have seen, was alarmed at the number of assumptions made and raised his concerns at the time; later stating that the plan 'depended on the unbelievable notion that once the bridges were captured, XXX Corps' tanks could drive up this abominably narrow road, allowing no manoeuvrability – and then walk into Germany like a bride into a church'.[5] Gavin, commander of US 82nd Airborne Division, in his memoirs, raises the question of why the operation was so badly planned.[6] So what was wrong with the plan? Let's return to Browning's Headquarters and conduct a postmortem on the operation.

Browning's Headquarters, Nijmegen, Holland, 26th September 1944

A key problem for the Market side of the operation was that the airborne divisions were dispersed over a wide area. Furthest south, the US 101st had been tasked with a huge operational area with numerous tasks and scattered

---

[2] Middlebrook, *Arnhem 1944*, 19.
[3] Ritchie, *Arnhem*, 147.
[4] Ibid., 256-257.
[5] Ryan, *A Bridge Too Far*, 116.
[6] Bennett, *A Magnificent Disaster*, 197.

landing and drop zones. Maxwell Taylor, its commander, had protested to Brereton and had got this changed.[7] The problem he faced was endemic across the operation, there was a fundamental tension between the need to take objectives outside of the area where the airborne divisions needed to hold landing grounds to facilitate later drops. This was exactly the problem that British 1st Airborne faced at Arnhem with the need to hold the landing and drop zones, keep the road to Arnhem open and hold the bridges.[8] US 82nd Airborne Division had been similarly stretched with the need to hold the Groesbeek Heights and take bridges in the Grave and Nijmegen area. The division's Chief of Staff, Colonel Robert Weinecke, protested beforehand that the task required two divisions.[9] Both Browning and Gavin bear responsibility for this overcommitment to multiple objectives.

Browning's mistake had been his insistence on prioritising the taking and holding of the Groesbeek Heights (to the east of Nijmegen) over the seizing of the bridges. The problem had been that the Heights needed to be held to block anticipated German counterattacks from the Reichswald Forest to the east on the border with Germany, meaning that the bridges were de-prioritised. Browning was on the hook for this decision. It was he who had written it into his operational instruction on 13th September, reiterating it (vehemently in Gavin's recollection) at a conference the next day.[10] The problem, as the operation unfolded, was that XXX Corps' advance had been severely held up due to the need to engage in an extended fight for the Nijmegen bridge. The failure to prioritise the capture of bridges was, however, not restricted to Nijmegen. Failure to drop British 1st Airborne Division close enough to their bridges further north and prioritising their capture had also caused problems at Arnhem. Leaving the bridges for capture later was, probably, fatal to the whole operation. The location of the landing zones and the prioritisation of objectives were decisions in which he had also been involved.

Some of the responsibility for the problems with the plan must rest with Montgomery. He was responsible for setting the broad parameters for the operation at Army Group level, assisted by Dempsey and, crucially, Browning. Montgomery was not fully conversant with airborne operations and the difficulties inherent in them. Previous experience with airborne

---

[7] Ritchie, *Arnhem*, 117.
[8] Bennett, *A Magnificent Disaster*, 228.
[9] Ryan, *A Bridge Too Far*, 89.
[10] Powell, *The Devil's Birthday*, 75.

operations had shown that they were very risky affairs, and Browning had allowed the risks to accumulate in Market Garden. He should, as the airborne advisor, have been more robust in advising Montgomery about the risks.[11] Although he would most likely not admit it, this was probably due to his lack of experience.

Browning's inexperience

Despite being intimately involved in the development of British Airborne Forces, Browning by September 1944, still lacked operational command experience. Ritchie argues this lack of experience meant he was unable to fully comprehend the dynamics and limitations of airborne operations and that, because of this, he should share responsibility for the problems with the D-Day drops and the problematical, but thankfully cancelled, Wild Cats operation which would have followed it in the Caen area. Gavin even contemporaneously recorded in his diary that Browning 'unquestionably lacks the standing, influence and judgement that comes from proper troop experience'.[12] This inexperience was, most likely evident in his insistence on prioritising the Groesbeek Heights but emerged in other ways. Most famously, his inability to understand the vulnerability of airborne forces can be seen in his misplaced assurance to Montgomery that 1st Airborne could hold on at Arnhem for four, not two days.[13]

Browning's inexperience and inability to fully understand the problems of airborne warfare also emerged during the battle itself. When the full reality of 1st Airborne's plight at Arnhem began to emerge, Browning was unable to grasp the sense of urgency required. Hakewill-Smith, who's 52nd Lowland Division was meant to be flown in at the end of the operation, wrote to Browning offering to commit his command early to the battle at Arnhem. Browning declined the offer and replied in a wholly unwarranted optimistic manner. On another occasion, Browning again did not seem to grasp the urgency of the situation when Roy Urquhart's Chief of Staff, Charles Mackenzie, swam the Rhine at Arnhem on the night of 22nd September to personally brief Browning about the desperate condition of the division.[14] This all points to a lack of experience that led to either Browning's lack of

---

1 Ritchie, *Arnhem*, 147-148.
12 Ritchie, *Arnhem*, 100.
13 TNA, CAB 106/1133, official historian's notes of an interview with Lieutenant General Sir Frederick Browning, 7 October 1954.
14 Mead, *General 'Boy'*, 136-139.

involvement in the situation, his inappropriate interference or ignoring advice of more experienced subordinates.

In summary, Browning's inexperience was an issue that caused problems, there were also, as with Montgomery, several cognitive biases that would lead him to commit some lapses of judgement. We shall look at these next.

Browning's Biases

Cognitive Dissonance

We saw in our discussion of Montgomery's biases how holding a strong opinion on a subject can overly influence your thinking, making it difficult to alter or change that view. We discussed the psychological phenomenon of Cognitive Dissonance, the state of mind or process where information that challenges a firmly held view needs to be ignored or negated in some way to keep the opinion intact. We have already discussed Browning's handling of the intelligence for Market Garden. His actions suggest that he too was experiencing a state of Cognitive Dissonance; he needed the operation to go ahead but was receiving increasingly worrying intelligence reports that challenged the feasibility of the operation. We do not need to repeat our discussion on Cognitive Dissonance here, but it is worth exploring a similar mechanism, Confirmation Bias.

Confirmation Bias is defined as the tendency to search for, notice, attend to, and process information that agrees with or confirms a closely held idea or hypothesis. Conversely, information that does not fit with that idea or challenges it in some way is not properly attended to. Information that supports a firmly held idea is more readily accepted and given less scrutiny whereas challenging information needs to be dealt with in some way.[15] Confirmation Bias is, therefore, very similar to Cognitive Dissonance, especially in terms of the mechanism that deals with difficult information. Confirmation Bias was demonstrated in a 1998 study conducted by two American psychologists, Simons and Levin. In this study, participants were stopped by one of the experimenters and asked for directions to a building. Halfway through the interaction, a door was carried between the participant and the experimenter who used the cover to swap with

---

15 Plous, *The Psychology of Judgment and Decision Making*, 231-234.

the other experimenter who carried on the conversation as if nothing had happened. When questioned later, just over half of the subjects did not notice the change.[16] The discrepant information, the new person in front of them, was ignored. This study clearly showed that we tend to accept information that supports and ignore information that challenges a previously held idea. This tendency can be seen in Browning's handling of the Market Garden intelligence.

By early September, there was a solid body of intelligence indicating the Wehrmacht was reinforcing the Market Garden operational area; Comet was cancelled due to the tougher resistance now being experienced by Second Army. This tougher German resistance was clearly stated in the Intelligence Appreciation for Comet but not included in the summary for Market Garden. This has led some commentators to ask whether knowledge of increased German opposition was deliberately suppressed.[17] More specifically, Harvey claims Browning did indeed omit details of Panzers in the Arnhem area.[18] If he did, this would be the active process of Cognitive Dissonance in action. Others lay similar accusations at Browning's feet.

Buckingham asserts that Browning was responsible for downplaying the German Panzer strength at Arnhem by withholding this information from the intelligence summaries handed down from Corps level to 1st Airborne Division. He cites Brian Urquhart's account of his own dismissal on medical grounds (under threat of court martial) and couples this with the absence of Panzers in summaries subsequently sent down to the division as further evidence of this.[19] He goes on to suggest Browning also suppressed the intelligence that indicated units in the Arnhem area were SS troops. Both omissions, led Roy Urquhart and Gerald Lathbury (commanding 1st Parachute Brigade) to prepare their plans assuming that they would be facing low-grade troops.[20] John Frost, (commanding 2nd Parachute Battalion) would later state that he would have taken more Anti-Tank weapons if fully aware of the threat.[21] Browning's headquarters issued the intelligence summary on 13th September that described armoured strength in the operational area as 'probably not more than 50-100 tanks, mostly Mark IV' and that troops

---

16 Chabris & Simons, *The Invisible Gorilla*, 59-60.
17 Powell, *The Devil's Birthday*, 46.
18 Harvey, *Arnhem*, 34.
19 Buckingham, *Arnhem 1944*, 75-76.
20 Ibid., 95.
21 Middlebrook, *Arnhem 1944*, 67.

in the area were 'not numerous and many are of low category'.[22] In support of this, Roy Urquhart states that Browning told him that his division would face opposition consisting of a 'Brigade Group, plus a few tanks'.[23] Putting all of this together paints a compelling picture of Browning's guilt. The issue, however, is not quite as clear cut.

A lot of the intelligence about the German (armoured) strength in the operational area was derived from Ultra, which was a very closely guarded secret and Browning, as a Corps Commander, was not read into this intelligence.[24] He did receive the limited Photo Reconnaissance intelligence provided by Brian Urquhart,[25] as well as intelligence summaries passed on from Twenty-First Army Group and Second Army, but, as illustrated, these tended to downplay the threat.[26] Powell refutes the notion that Browning deliberately suppressed intelligence and argues that he was 'in no way the man to endanger his troops'.[27] John Buckley adopts perhaps the most rounded view on the issue. He suggests Browning did not deliberately suppress the intelligence but instead did not realise its potential significance.[28] Sosabowski broadly agrees with this point, stating in his memoirs Browning was too optimistic.[29] If this is the case, we would argue that Browning's desire to see the operation go ahead meant he had bought into the picture of light resistance and so, suffering from Confirmation Bias, was prepared to accept the low threat intelligence he was privy to. There was another reason why he was prepared to view the situation this way. He was also, as we have seen, keen to see the investment in airborne forces pay off, and to get into action himself; he was, therefore, perhaps too committed to the operation.

Endowment Effect

Chapter Six explored the pressures that Browning was under at the time of Market Garden. In particular, we examined Cialdini's social influence factor of Commitment and Consistency, whereby once committed to something you are more likely to feel obliged to follow it through. We can see this factor

---

22 Powell, *The Devil's Birthday*, 45.
23 Urquhart, *Arnhem*, 9.
24 Powell, *The Devil's Birthday*, 47.
25 Bennett, *A Magnificent Disaster*, 199.
26 Hibbert, *The Battle of Arnhem*, 37.
27 Powell, *The Devil's Birthday*, 47.
28 Buckley & Preston-Hough, *Operation Market Garden*, 216.
29 Sosabowski, *Freely I Served*, 194.

helping to drive Browning's commitment to seeing airborne forces used in a strategic role, and we can see it in a specific cognitive bias called the Endowment Effect.

The Endownment Effect specifically refers to the process by which you place a high value on an item because you 'own' it. More broadly, it means that if you have invested time, effort or money into something, such as an idea or a project, you 'endow' it with a very high value. Studies have shown that the Endowment Effect means not only is an item more highly valued, but individuals are more reluctant to let the item go.[30]

As we have seen, by 10th September, when Operation Comet was replaced by Market Garden there had been considerable investment in the development of airborne forces in the Allied camp, both physically and psychologically. Browning had been central to this effort for three years. Most recently, FAAA had been created on 8th August 1944; it was composed of two Corps and approximately 50,000 men and formed the Allied strategic reserve, but it was 'coins burning holes in SHAEF's pocket'.[31] As John Peaty points out, having created this elite, the airborne forces saw a disproportionately shorter time in action than regular infantry units and even less time in the role for which it had been created.[32] First Airborne Division is a very good example. As of September 1944, the division had seen only 17 days of action and had not been employed since September 1943.[33] Browning was, we would argue, suffering from the Endowment Effect and feeling the imperative to see the (British) airborne forces employed in a decisive, strategic role for which he would get the credit. If it didn't, he would miss out on his last chance for an operational command.

Loss Aversion

Daniel Kahneman and Amos Tversky's research into cognitive biases posits that you will perceive real or potential losses as more psychologically impactful than an equivalent gain. For example, you will feel comparatively more negative emotion about losing £100 than you would feel positive emotion about gaining £100. This is known as Loss Aversion.[34] Browning's ambition and need for an operational command meant that he was facing a

---

30 Kahneman, *Thinking, Fast and Slow*, 289-299.
31 Powell, *The Devil's Birthday*, 11-28.
32 Peaty, *Operation Market Garden*, 72-73.
33 TNA, WO 205/873, Allied Airborne Operations in Holland, September to October 1944.
34 Hardman, *Judgement and Decision Making*, 66-67.

potential loss if Market Garden was cancelled or even postponed. He also needed to be seen to be in command of the operation; this brings us to another controversial episode.

Browning chose to command the operation from the central location of Nijmegen. To do this, he needed to insert his Corps Headquarters despite his staff having no operational experience or proper establishment.[35] This lift required 38 gliders which he took from 1st Airborne Division, which meant Urquhart could only bring in half of his 1st Airlanding Brigade on the first day.[36] This, as we shall see in the next part of the book, would have a serious impact on the division's ability to generate combat power during the crucial first 24 hours of the operation. The key problem with this decision was that the operation did not require the presence of the Corps Headquarters.[37] The staff had little to do and made no real difference to the conduct of the operation.[38] This was quite evident at the time, to the point that the staff at 1st Parachute Brigade at Arnhem wrote a poem, ridiculing Browning's staff. One verse illustrates the point nicely: 'And what they Command when they get there, If Command is their ultimate goal, Is a matter it's hard to conjecture, We ourselves and a possible Pole. But it's better than drawing the dole!'[39] Browning's biographer makes an attempt to defend his subject, but it is hard to disagree with the view, expressed by most commentators, that bringing in his own headquarters was an unnecessary act of vanity and he did it so that he could be seen to be in command of the operation.[40] This would have been the crowning moment of his long association with the airborne forces, cementing his reputation as the expert on the subject.

Browning's involvement with the development of the airborne capability also meant he was in a good position to draw on the lessons learned from previous operations. This can be a good thing if the correct lessons are applied, it can also have a negative impact if the past experiences are not applied in the correct manner.

---

35 Mead, *General 'Boy'*, 105.
36 Buckingham, *Arnhem 1944*, 91-92.
37 Middlebrook, *Arnhem 1944*, 11.
38 Powell, *The Devil's Birthday*, 241.
39 Middlebrook, *Arnhem 1944*, 12.
40 Buckingham, *Arnhem 1944*, 91.

Failure to Learn

By 1944, the Allies had had the opportunity to learn lessons from previous operations, both their own and that of the Germans. The problem was that the wrong lessons were learnt. The Allies had been inspired to develop their own airborne capability by the effective German operations in 1940.[41] These operations were successful mainly due to being generally small scale (Crete apart), undertaken against limited opposition, were relieved by ground troops quickly and preceded by extensive rehearsal.[42] Even then, these operations tended to not achieve absolute mission success.[43] German experience at Crete in May 1941 demonstrated that large scale airborne operations were expensive in terms of both lives and equipment, but the Allies drew the wrong lesson from this operation, viewing it as a successful strategic use of the capability.[44] The Germans saw things differently, the heavy casualties sustained meant Hitler banned further large scale airborne operations.

Allied deployment of airborne forces in the Mediterranean again drew the wrong lessons from their use. Their experience in North Africa highlighted the key issues that the German operations had raised, the importance of good command and control, the necessity for rapid link-up with ground forces, enough lead time to facilitate proper planning and good air lifts to deliver a close concentration in the drop.[45] These emphases do not seem to have been taken on board by the Allies as the same mistakes were made during their invasion of Sicily.[46] The other key lesson from the experience at Sicily, was the need to match the air plan to the ability of the crews undertaking it.[47] The night time insertion delivered by undertrained air crews meant the drop was disastrously dispersed, with a number of gliders landing in the sea. This problem was repeated a year later during D-Day, where the night drops meant the two American airborne divisions, in particular, were widely scattered.[48] This experience should have reinforced the four key lessons highlighted previously, and also that the reaction of the enemy needed to

---

41 Powell, *The Devil's Birthday*, 246.
42 Ritchie, *Arnhem*, 24.
43 Ibid., 36.
44 Ibid., 36-38.
45 Ritchie, *Learning to Lose?*, 22.
46 Ritchie, *Arnhem*, 53.
47 Ritchie, *Learning to Lose?*, 24.
48 Warren, *Airborne Operations*, p.61.

be factored in to planning more thoroughly.[49] Ritchie points out that many of the airborne objectives on D-Day were not achieved but this was masked by the overall success of the landings.[50] The right lessons from the D-Day airborne experience were not learnt.[51]

Browning was intimately involved in these operations as airborne advisor and had the opportunity to draw the correct lessons from this experience. To summarise, the key points that this experience showed were the effectiveness of small scale and not large operations, and that complete mission success was not achievable. There was also the need for plenty of lead time for planning and rehearsal, the need for limited opposition and quick relief by ground forces, which by implication meant not dropping too deep into enemy territory. Previous experience also highlighted the importance of good intelligence on enemy dispositions and the nature of the landing zones, the need to drop close to the objectives and the need to factor in the reaction of the opposing forces. The need for good command and control, under one organisation, to match the air plan to crew capabilities and the problem that night drops would probably cause dispersal were also important. As we have seen, apart from switching to a daytime drop, these lessons were not absorbed and the errors were repeated, and on a grander scale in Market Garden. The operation broke most of the rules. Ritchie argues that the thinking about the use of airborne forces was unrealistic.[52] Buckley agrees, arguing that the Allies overestimated the efficacy of airborne operations.[53] This was partly due to the misapplication of the experiences the Allies had gained from such operations; Browning was central to the process as the key airborne advisor.

Conclusion

Market Garden was based on flawed assumptions about the efficacy of airborne operations, on 'massaged' intelligence and acceptance of multiple and cumulative risks.[54] As shown, there is good evidence to suggest Browning's involvement in all three of these issues, so in part he was responsible for the errors ultimately leading to the failure of Market

---

49 Ritchie, *Learning to Lose?*, 26.
50 Ritchie, *Arnhem*, 65.
51 Ibid., 74.
52 Ibid., 91.
53 Buckley & Preston-Hough, *Operation Market Garden*, 210.
54 Ritchie, *Arnhem*, 256.

Garden. Regardless of his culpability, Browning seems to have taken steps to dissociate himself from (the failure of) the operation.

Browning has been criticised for not meeting the 1st Airborne evacuees when they were brought back over the Rhine. He did not really engage with Roy Urquhart when he made it to Corps Headquarters, Urquhart later describing their meeting as 'totally inadequate'. Browning was also rather absent on the night following the evacuation of 1st Airborne while fighting was still happening at different points along the corridor.[55] When Brian Urquhart asked Browning for a posting away from Corps Headquarters, the latter asked him to keep their various disagreements to himself.[56] Finally, perhaps worst of all, Browning (and Montgomery) later tried to pin the blame on the Poles, Sosabowski in particular. Browning wrote a bad report on Sosabowski and essentially got him removed from his command.[57] Some might argue that there was a certain poetic justice in the fact that Market Garden, rather than enhancing his prospects, effectively killed off Browning's military career.[58] In the British way of dealing with these issues, he was knighted but moved from both his posts and sent out to the Far East as Chief of Staff to the supreme commander there.[59]

For the reasons discussed above, several commentators see Browning as the malign influence behind Market Garden; Buckingham certainly portrays him as the pantomime villain (as does the film). Mead, his biographer, makes some attempt to defend him but does accept that he was perhaps guilty of ignoring, though not suppressing the worrying intelligence and could have done more to challenge the air plan and the location of the landing and drop zones.[60] It is probably a fair assessment. Faced with situational pressures to proceed with the operation and driven by his own ambition, he was probably too optimistic, overestimating the airborne forces capabilities and underestimating those of the enemy, and thus ultimately failed to engage in enough critical appraisal of the viability of the operation. As with Montgomery in the prior section, we will conclude this part of the book by exploring in the next chapter how a Structured Analytical Technique might have helped Browning take a more robust critical approach.

---

55 Mead, *General 'Boy'*, 146-147.
56 Ibid., 150.
57 Ibid., 165.
58 Ibid., 235.
59 Ibid., 167.
60 Mead, *General 'Boy'*, 154-155.

10
Feasibility

Introduction

As we have seen in the previous chapters, once the operation was handed over to Browning and FAAA, there were clear indications the mission would be challenging to say the least. There were clear risks, to the extent that the feasibility of the operation could be open to question. Some of these risks were ignored or downplayed in the airborne community's eagerness to get into action, and it is clear from the previous analysis that key decision-makers were perhaps not thinking as critically as they could have done. This chapter uses another Structured Analytical Technique, the Cone of Plausibility, as a framework for exploring how critical thinking may have helped shape Browning's thinking at the time, especially about the feasibility of the operation.[1]

Feasibility

There were several planning considerations or factors, both internal and external, that would affect the feasibility of the operation and potentially impact its outcome. It will be useful to recap these before working through the Cone of Plausibility exercise.

Reaction of the Germans (stiffening resistance)

Previous chapters have highlighted the existence of intelligence to suggest German resistance was stiffening and that perhaps reaction to the landings might not be as weak as initially expected. Indeed, Browning was informed on 10th September about the presence of II SS Panzer Corps regrouping in the

---

[1] Heuer & Pherson, *Structured Analytical Techniques for Intelligence Analysis*, 141-142.

area.² In contrast, we have already noted the operational instruction issued to I Airborne Corps stating that 'the troops ... are low category' and German 'armoured strength is 50-100 tanks, mainly Mark IVs.'³ The intelligence picture was, therefore, at least at Corps level, unclear, the reaction of the Germans was clearly a variable that would affect the outcome of the operation.

Rate of Advance for XXX Corps

The operational concept called for the relief of Arnhem within 48 hours, thus requiring a rapid advance by the Garden forces. On paper, XXX Corps under Brian Horrocks was a strong formation with the Guards Armoured Division (Adair), 43rd Division (Thomas), 50th Division (Graham) and 8th Armoured Brigade under command. The problem, as previously noted, was XXX Corps would be required to advance along an exposed single road, which was raised in places. The nature of the terrain also meant it would be difficult to manoeuvre off the road should problems arise. XXX Corps' ability to maintain the rate of advance required would be another factor that would need to be considered.

Narrow Salient for XXX Corps

Not only was XXX Corps expected to advance along a single road, but its whole axis of advance was also along a narrow salient. With German forces sitting on either side of the salient, the exposed flanks meant there was a risk of the road being cut, thus halting XXX Corps advance. The advance would also use VIII Corps (O'Connor) and XII Corps (Ritchie) as flank protection, but these formations would be required to keep up with the pace of XXX Corps and, be in place in time for the start of the operation. The narrowness of the salient and possibility of the road being cut were clearly issues that could threaten the operation.

Ability of Airborne Forces to Generate Combat Power

Another factor to consider was the need for the airborne forces to quickly generate sufficient combat power to seize and hold bridges at each target location. As we have seen, the Market forces available were substantial,

---

2 Buckingham, *Arnhem 1944*, 253.
3 Powell, *The Devil's Birthday*, 45.

including 101st US Airborne Division (Taylor), 82nd US Airborne Division (Gavin) and 1st British Airborne Division (Urquhart). As discussed at length above, the choice of daylight drops, and limitations on aircraft, meant these divisions would be inserted over three days in some cases. In the middle, the need to hold the drop zones and control the Groesbeek Heights (US 82nd) meant that limited troops would be committed to seizing the bridges.[4] While at the tip of the operation the example of British 1st Airborne is illustrative. Out of a division of around 12,000 men, only one battalion (650 men) was tasked with seizing Arnhem Bridge.[5] This problem was compounded by the distance from the landing and drop zones and the requirement to take multiple objectives. The ability of all three airborne divisions to quickly seize and hold the bridges assigned to them was clearly another variable that would need to be considered.

Need to Seize Interlocking Bridges

Inherent in the plan, was the need to seize interlocking bridges. The issue here, clearly, was the loss of one bridge would delay the advance of XXX Corps and thus disrupt the timetable for the operation. The bridges were defended and rigged for demolition, so even with the employment of *coup de main* parties, landing very close to the objective at night, there would have been a risk of losing a bridge or bridges. As we have seen, use of *coup de main* parties were ruled out, and the operation involved a daylight drop. Losing one or more bridges was clearly a real possibility.

Weather

The final factor or variable to consider is the weather. The operation was taking place in mid to late September in Northwest Europe. There was a risk of fog in England that would prevent reinforcement and resupply, and cloud cover in Holland that would prevent Close Air Support. The need for a period of fine weather was clearly a variable that would affect the operation. Now that we have identified the different factors that might affect the outcome of Market Garden, we can turn to the Cone of Plausibility exercise to examine its feasibility.

---

4 Ritchie, *Arnhem*, 117.
5 TNA, WO 171/393, 1 Para Bde OO No. 1, 1st Airborne Division War Diary, September – December 1944.

Cone of Plausibility

The Cone of Plausibility exercise was designed to create different scenarios to illustrate how a situation might unfold. It can, therefore, be used to create alternative perspectives on different possible future outcomes and so is useful to help us explore the feasibility of Operation Market Garden. The basic idea is to use an understanding of the different factors or drivers (typically drawn from the Political, Economic, Social, Technological, Legal and Environmental (PESTLE) framework) that might shape how a situation develops over time. The aim is to generate up to six (although typically four) different scenarios. The different variations are the Mainline Scenario, which describes how the situation will most likely unfold, Good Case or how the situation will unfold if the drivers develop in a reasonably positive manner and Best Case, which is the optimal outcome where all or most of the factors develop positively. Conversely, the Bad Case Scenario describes how the situation will unfold if the drivers develop in a reasonably negative manner and Worst Case, or the least preferred outcome where all or most of the factors develop negatively. Finally, there is a Wildcard Scenario which outlines how a situation might unfold if an unlikely (black swan) event occurs. These different scenarios can then form the basis for planning or, as in our case, evaluating the feasibility or the likelihood of success. The exercise follows several steps.

The first stage is to define the problem by clearly stating the problem to be considered with as much specificity as possible. Ideally, a timescale should be attached to the problem statement. In our case, the question is how feasible is Operation Market Garden? Stage two is to brainstorm a list of factors or drivers that could shape how the problem situation might develop over the given timescale. These possible drivers should be stated in neutral terms such as 'rate of advance' and not 'slow' rate of advance. The next stage is to develop the different scenarios by working through the following steps. Firstly, start with the Mainline or most likely scenario and select the four to six drivers that are most likely to shape the way the situation will develop. We have already identified the relevant factors for our analysis: reaction of German forces; rate of advance for XXX Corps; narrow salient for XXX Corps; ability of airborne forces to concentrate and generate combat power (especially 1st Airborne); need to seize interlocking bridges; finally, the weather. Other factors could be used, but for the purposes of this exercise, we will run with these considerations.

Factor	Driver	Assumption	Impact

Overview:

Conclusion:

Table 10.1 Scenario Analysis Template

Feasibility

Driver	Assumption	Impact
Reaction of Germans:	Substantial but battered forces in the airborne area, quick reactions lead to blocking forces in place.	Blocking forces prevent rapid seizure of all bridges, necessitates greater concentration of force.
Rate of advance for XXX Corps.	Sporadic pockets of German resistance, creates multiple instances where deliberate attacks are required.	Halts to clear route slows rate of XXX Corps advance; imposes 72-hour delay.
Narrow salient for XXX Corps.	XXX Corps outpaces flanks, salient cut on two or three occasions, requires diversion of resources to re-open.	Halts to counterattack and clear route slows rate of XXX Corps advance; imposes 48-hour delay.
Ability of airborne forces to generate combat power.	Cohesive drop, but unable to concentrate enough power to quickly overcome determined resistance.	Unable to fight through to all objectives in sufficient strength, Germans hold some bridges. Imposes 24-hour delay.
Need to seize interlocking bridges.	One or two bridges destroyed, requires bridging units to replace.	Halts to build bridges slows rate of XXX Corps advance; imposes 24-hour delay.
Weather.	Short period of fog in England delays second or third lift.	Limits airborne resources on ground, reduces combat power and restricts freedom of movement.

Overview:
Distances of landing and drop zones from bridges gives German units time to react and put blocking forces in place. Despite a cohesive daytime drop that facilitates rapid forming up, airborne units are unable, due to limited numbers, to concentrate sufficient combat power to fight through these blocking forces. Poor weather in England slows rate of reinforcement and re-supply which compounds the airborne lack of fighting power. This results in some objectives not being taken quickly and hampers airborne ability to hold on to ground. The need to coordinate airborne and ground units to undertake deliberate attacks on bridges slows XXX Corps' rate of advance. The use of a single lane road means sporadic but effective German resistance on XXX Corps' route blocks the highway on multiple occasions, with XXX Corps unable to manoeuvre around the obstacles; this significantly further slows XXX Corps' advance. This rate of advance is further slowed by the need to replace destroyed bridges and repel German attacks on the XXX Corps' exposed flanks in the salient.

Conclusion:
Multiple delays slow XXX Corps' rate of advance, it arrives at Arnhem after 7 days. This is too late as 1st Airborne has been unable to take and hold Arnhem bridge and has been defeated in detail. XXX Corps are unable to force a crossing of the Rhine.

Table 10.2 Mainline Scenario

The next step is to describe how each driver will unfold within the context of the type of scenario, positively or negatively (called Assumptions in Table 10.1). Each should be a one sentence description of how that driver will develop over the set time frame. The next step is to take each Assumption and define the Impact the development will have on the problem situation. This should again be a brief, one sentence description.

The final stage is to synthesise the different impacts into one holistic description of how the situation will develop. This overview should incorporate all the impacts; it can be useful to develop it in chronological order where each impact affects others in a sequence. Based on this overview, it should be possible to develop an overall conclusion. The nature of the conclusion will depend on the context of the problem situation.

An example Mainline Scenario for Market Garden following these steps is outlined in Table 10.2.

As can be seen with the interpretations embodied above, the most likely outcome is mission failure. XXX Corps' rate of advance is just too slow, and it takes too long to arrive at the ultimate prize, Arnhem. The airborne's inability to concentrate sufficient combat power to take and crucially hold all the objectives means the bridges are not being held when XXX Corps eventually arrive. This is especially the case with Arnhem.

It is, of course, important to be careful about hindsight bias and 'situating the estimate' with what we know happened. In terms of the choice of factors, it should be noted, however, that a number of these issues (German reaction, weakening of airborne combat power, loss of surprise, and the problem of using a single road) were concerns voiced at the time. In terms of the assumptions, previous experience of fighting the Germans indicated that they had the ability to react, and intelligence indicated the German units in the area were specially trained in anti-Airborne tactics.[6] Airborne doctrine and previous experience also indicated problems with not dropping close to objectives in sufficient strength to seize them. Finally, in terms of impacts, a more measured assessment could have been achieved by using specialists to calculate each delay. We have used rough estimates.

On a positive note, the assessment does point to areas that needed to be addressed; it at least identifies the risks that are being tolerated if the whole thing was a *fait accompli,* for example the selection of the landing and drop zones, or, that in risk management terms are being 'treated' to help

---

5 Kershaw, *It Never Snows in September,* 44.

improve the chances of success. For example, the delays imposed by German resistance to 1st Airborne's advance to the bridges should have highlighted the vital importance of Close Air Support (CAS). A conscious recognition of its importance may have led to a re-examination of its (relative lack of) use during the operation.

Overall, based on this assessment, the operation was not feasible as it stood and would need all the chance factors to go its way for it to have a chance to succeed, this is where generating and exploring other scenarios might help to examine whether different circumstances or better luck might have led to mission success. Let's go back to our process.

Other scenarios can then be developed by repeating the analytical steps outlined above. In each instance the assumptions (the way in which the driver develops) are changed to reflect the type of scenario. For the Good Case scenario, we change roughly half of the assumptions so that they are positive in tone; these should be the drivers that are most likely to develop in a positive manner. We change all or most of the assumptions to be positive in tone for the Best Case. Conversely, for the Bad Case scenario, we change roughly half of the assumptions to be negative in tone; again, these should be the drivers that are most likely to develop in a negative manner. We also change all or most of the assumptions to be negative for the Worst Case. Finally, for the Wildcard Scenario, we change one or possibly two of the drivers to reflect a high impact-low probability event. In that instance, a new driver could be introduced to reflect such a black swan event.

It is best to use the same drivers for each scenario and change the Assumptions if possible; it is, however, possible to use different drivers if that makes sense. For each scenario, the impact linked to the changed assumption will be different as will the overview and hence the conclusion; this is essentially the key reason for conducting the exercise. For the purposes of this exercise, we will examine only the Good and Best-Case scenarios to examine the feasibility of the operation, just how much luck did the Allies need for the operation to be successful, how much of a gamble was it? Let's allow ourselves a little bit of luck and have a look at the Good Case scenario (the changes in Table 10.3 are in italics).

As can be seen, in the Good Case scenario, using the altered factors and assumptions, with some factors going the Allies' way, a degree of success is achieved. The operation succeeds in establishing a bridgehead across the Rhine, but the hard charging by XXX Corps and British 1st Airborne, and the prolonged defensive action of the latter leads to a very high casualty rate.

Driver	Assumption	Impact
Reaction of Germans:	Substantial but battered forces in the airborne area, quick reactions lead to blocking forces in place.	Blocking forces prevent rapid seizure of all bridges, necessitates greater concentration of force.
Rate of advance for XXX Corps.	*Pockets of German resistance, XXX Corps commanders take risks and show more drive when cleared.*	*Halts but XXX Corps moves at night, armour moves without infantry, 48-hour delay.*
Narrow salient for XXX Corps.	XXX Corps outpaces flanks, salient cut on two or three occasions, requires diversion of resources to re-open.	Halts to counterattack and clear route slows rate of XXX Corps advance; imposes 48-hour delay.
Ability of airborne forces to generate combat power.	*Cohesive drop, flexible tactics ensure airborne units can manoeuvre in strength.*	*Airborne able to fight through to all objectives in sufficient strength, key bridges seized.*
Need to seize interlocking bridges.	One or two bridges destroyed, requires bridging units to replace.	Halts to build bridges slows rate of XXX Corps advance; imposes 24-hour delay.
Weather.	*Good weather in England means second and third lifts go in on time.*	*After delay, airborne combat power is increased, bridges reinforced.*

Overview:
Distances of landing and drop zones from bridges gives German units time to react and put blocking forces in place. A cohesive daytime drop that facilitates rapid forming up, combined with flexible manoeuvre means airborne units can concentrate sufficient combat power to fight through these blocking forces. Good weather in England facilitates the rate of reinforcement and re-supply which increases airborne fighting power. This results in objectives being taken quickly and held in strength but with significant casualties incurred in the defence of the objectives. The use of a single lane road means sporadic but effective German resistance on XXX Corps' route blocks the highway on multiple occasions, with XXX Corps unable to manoeuvre around the obstacles; this considerably slows XXX Corps' advance. Their rate of advance is further slowed by the need to replace destroyed bridges and repel German attacks on XXX Corps' exposed flanks on the salient. An injection of urgency from senior XXX Commanders means doctrine is ignored with night moves undertaken and, in some instances, armour moving forward without infantry support. This mitigates the delays to some extent but with significant casualties.

Conclusion:
Multiple delays slow XXX Corps' rate of advance, it arrives at Arnhem after 5 days but has suffered significant casualties. British 1st Airborne holds Arnhem bridge long enough to be relieved but has also suffered significant casualties. XXX Corps is across the Rhine but unable to exploit the bridgehead. Operational but not strategic success; a pyrrhic victory.

Table 10.3 Good Case Scenario

Feasibility

Driver	Assumption	Impact
Reaction of Germans:	Weakened forces in the airborne area, unable to respond effectively and block movement.	Rapid seizure of all bridges, objectives held in strength.
Rate of advance for XXX Corps.	Pockets of German resistance, XXX Corps commanders take risks and show more drive when cleared.	Halts but XXX Corps moves at night, armour moves without infantry, limited delay to timetable (24 hours).
Narrow salient for XXX Corps.	Flanking Corps ready on time and keep pace with XXX Corps, integrity of salient is maintained.	XXX Corps advance unimpeded by flank attack, sticks to timetable.
Ability of airborne forces to generate combat power.	Cohesive drop, flexible tactics ensure airborne units can manoeuvre in strength.	Airborne able to fight through to all objectives in sufficient strength, key bridges seized.
Need to seize interlocking bridges.	Germans fail to demolish any bridges, all objectives seized.	XXX Corps advance moves across bridges unimpeded, sticks to timetable.
Weather.	Good weather in England means second and third lifts go in on time.	Airborne combat power is increased, bridges reinforced.

Overview:
Distances of landing and drop zones from bridges gives German units time to react and put blocking forces in place, but these forces are weak. A cohesive daytime drop that facilitates rapid forming up, combined with flexible manoeuvre means airborne units can concentrate sufficient combat power to fight through these blocking forces. Good weather in England facilitates rate of reinforcement and re-supply which increases airborne fighting power. This results in objectives being taken quickly and held in strength but with heavy casualties including in defence of the objectives. The rate of advance is facilitated by the successful seizing of all bridges and lack of serious threat to XXX Corps' exposed flanks in the salient. An injection of urgency from senior XXX Commanders means doctrine is ignored with night moves undertaken and, in some instances, armour moving forward without infantry support. This mitigates the sporadic German resistance on XXX Corps' route that blocks the highway on occasions. The rapidity of advance along the route means German forces have less time to react and rush reinforcements into the area, this creates better conditions for XXX Corps and further facilitates its advance.

Conclusion:
Limited delays do not appreciably slow XXX Corps' rate of advance. It arrives at Arnhem after 3 days. British 1st Airborne holds Arnhem bridge long enough to be relieved but suffers a high casualty rate. XXX Corps is across the Rhine and in a position to exploit the bridgehead. Operational and potentially strategic success but at the cost of sacrificing British 1st Airborne.

Table 10.4 Best Case Scenario

This rate is so high that XXX Corps is unable to exploit the bridgehead and so the strategic objective is not achieved.

We have chosen to change what we believe to be the most likely factors that could have had a positive impact. Luck in terms of the weather, a change in command approach with XXX Corps pushing on harder and British 1st Airborne not committing all the battalions from 1st Parachute Brigade to different routes and thus being able to push the whole formation down the lower road. Selecting other factors and assumptions would of course change the outcome and conclusion.

In summary, even in the Good Case scenario, it seems the operation is still not really feasible, only operational not strategic success is achieved. So, let's turn to the Best-Case Scenario in Table 10.4 and explore what might have happened if everything had gone according to plan.

As can be seen from Table 10.4 in the Best-Case Scenario, using these additionally altered factors and assumptions, with all the factors going the Allies' way, success is achieved but at a cost. The operation succeeds in establishing a bridgehead across the Rhine, with the rapidity of XXX Corps advance meaning it is, in theory at least, in good condition to exploit the situation. It is unclear whether Eisenhower would have chosen to support this move and indeed, whether Montgomery's Twenty-First Army Group was strong enough to effectively undertake the subsequent action. British 1st Airborne's extended defensive action leads to a high casualty rate, rendering it unfit for further service. It has, essentially been sacrificed in the operation.

In summary, Market Garden is a success in the Best-Case Scenario, but everything must go the Allies' way and at the operational level things needed to be done differently. Of course, the probabilities of the chance factors working out for the Allies are small. The likelihood of those involved in the operation dramatically changing their approach and doing things differently at such short notice is also small. So, was there a chance that Market Garden would work as planned? Yes, but a small one, only in a Best-Case Scenario. This assessment rather underscores the view that when examined at an operational level, Market Garden was not feasible. Those involved, including Browning specifically, were able to raise firmer objections and/or adjust the plan. They, and he, did not.

Operation Market Garden is, then, going ahead as planned. It is now time to move down to the 'tactical' level. Major-General Roy Urquhart and

British 1st Airborne Division will be dropping on Arnhem, so let's explore his decision-making.

Part Three

Reproduced with kind permission of Airborne Assault Museum.

Urquhart

11

Urquhart's Observation – Conformity

'Most of them sat nonchalantly with legs crossed, looking rather bored'.[1]

Introduction

Urquhart's Headquarters, Grantham, England, 12th September 1944

The last two days had been a whirlwind of activity for Roy Urquhart. He and his staff had been busy planning 1st Airborne's role in Market Garden. Late on the 12th, he had just finished briefing his senior officers at an Orders Group at Divisional Headquarters at Grantham. He had been informed by Lieutenant-General Browning on 10th September that his division would get the prize – Arnhem. He had immediately set to work, but it had been something of a challenge. Although a professional soldier with considerable operational experience, he was an infantryman and not by background an airborne officer. He felt this lack of airborne experience acutely and was in a rather difficult situation.

Urquhart's situation

To understand how Urquhart responded to the situation he faced, it will be useful to examine his background and prior service; an overview of the current situation, and the task demands placed upon him. Urquhart received his orders on 10th September from Browning as the deputy commander of FAAA to begin planning 1st Airborne's role in Operation Market Garden.[2] Added to the challenging task posed by this order would have been other

---

1 Sosabowski, *Freely I Served*, 145.
2 WO 171/393 HQ 1st Airborne Division War Diary, September- December 1944.

pressures that existed at the time; the social situation, various organisational issues and the time frame within which the operation had to be planned. Urquhart, therefore in addressing the situation (the first stage of the OODA Loop) was required to face both task-related and situational pressures. We will again, therefore, examine these situational pressures through the lens of Cialdini's social influence factors. Let's return to Urquhart's Headquarters.

Urquhart's Headquarters, Grantham, England, 12th September 1944

Urquhart had received the operational instruction to seize the bridges at Arnhem on 10th September, it was, therefore, a lawful order.[3] He was also aware that the operation was supported all along the chain of command, from his immediate superior (Browning), through the commander of his parent formation (Brereton) and senior commanders in theatre (Eisenhower and Montgomery), all the way to governmental level (in both London and Washington). He knew that if he had any reservations, it would be difficult to resist and question the order; and he did have concerns, there were clearly problems with his part of the operation, but he felt that, as he would later state in his memoirs, he was 'left with no choice'.[4]

Urquhart was in something of a weak position. Firstly, although he was an experienced, professional soldier, he had no airborne experience. He had been commissioned into the Highland Light Infantry in 1921 and had served in different regimental and staff posts before the war, attending Staff College at Camberley between 1935-37 before serving in India for two years. He had been posted back to the UK at the outbreak of the war, taking up a staff appointment at 3rd Division, where he had met Montgomery. He had quickly moved on to command 2nd Battalion of the Duke of Cornwall's Light Infantry (DCLI), with whom he remained (in the UK) for two years. Most recently, he had served in the North African, Sicilian and Italian campaigns in staff and command roles.[5] He had been the Chief of Staff for 51st Highland Division and commanded 231st Independent Infantry Brigade Group between 1942 and 1943.[6] He could, therefore, draw on solid experience as an infantry officer but was now, commanding an airborne division. He would later note in his autobiography: 'I was very much aware of my lack of experience of airborne operations'.

---

3 WO 171/393 HQ 1st Airborne Division War Diary, September- December 1944.
4 Urquhart, *Arnhem*, 7.
5 Baynes, *Urquhart of Arnhem*, 1-66.
6 Harclerode, *Arnhem: A Tragedy of Errors*, 54.

The situation had been compounded by his reception upon taking up his command when (as he would later note that) he was aware of 'being looked over closely with ill-concealed reservations in some quarters'.[7]

Urquhart had also not had long (about eight months) to establish himself in his new command; he had also only been recently promoted. When he took over command of the division in January 1944, in terms of seniority of rank, he was a substantive Lieutenant-Colonel, a temporary Brigadier and only an acting Major-General.[8] This was not unusual for the British Army during the Second World War, but this does not detract from the fact that this was the situation facing Urquhart. Added to this, Eric Down, his predecessor in commanding the division had been removed by Browning (to establish an airborne division in India) because he was too experienced and critical of his superior. Gerald Lathbury, who commanded 1st Parachute Brigade within Urquhart's new command, was an obvious choice to have taken command but had been overlooked.[9] All of this served to weaken Urquhart's sense of authority and his capacity to challenge the order he had been given.

Urquhart's position

Urquhart's statements, quoted above, need to be treated with some caution as they are drawn from his autobiography and may be attempts at defending himself against criticisms of his actions by emphasising his lack of airborne experience. The point is partly substantiated by John Frost, who, as Commanding Officer of 2nd Parachute Battalion, 1st Parachute Brigade, was one of Urquhart's subordinates, refers to the difficulties of bringing in someone from outside the airborne community.[10] Frost's memoir is that of another participant in the operation but does provide some cross-validation. It seems clear that Urquhart, being somewhat of an airborne 'outsider' would have found it very difficult to argue his case with unquestioned credibility.

Some commentators suggest that Urquhart's inexperience, especially in airborne matters, was the reason why he was accepted by Browning as it meant the latter could exert more control over him.[11] Regardless of whether it is true, we would suggest that he was more likely to acquiesce than his predecessor, Eric Down. This view has more validity when we

---

7 Urquhart, *Arnhem*, 14-15.
8 Roy Urquhart, *Rising Eighty*, Urquhart Papers, Imperial War Museum, 102.
9 Buckingham, *Arnhem 1944*, 26-31.
10 Frost, *A Drop Too Many*, 195.
11 Buckingham, *Arnhem 1944*, 30.

consider Urquhart's character. The various pen pictures painted about him portray Urquhart as straightforward and low key. General Sosabowski described Urquhart as 'pleasant and easy to work with' and a 'kind man'.[12] 'Kind' was the word that David Niven used to describe Urquhart when he knew him earlier in his career.[13] These sentiments are supported by comments made by the Adjutant and Padre during his time commanding the Duke of Cornwall's Light Infantry.[14]

Urquhart, due to situational and dispositional factors, therefore lacked a firm basis from which to pose any challenge should he have wished to do so. He was in a difficult position as the newly appointed General Officer Commanding the division and was on a very steep learning curve with little time available. This situation was not helped by him being debilitated by a re-occurrence of Malaria in early 1944, which took him out of action for a few weeks.[15] This inevitably had a negative impact on his ability to take control of the division and drive through the formation level training that it needed.[16]

There is, therefore, evidence to suggest that Urquhart was at a disadvantage that would have made it difficult for him to resist the pressures of authority inherent in the situation. Given the circumstances, he might have looked to his peers or his staff officers for support in questioning the viability of the operation, or at least to gauge as to whether it was a sound concept; this is where the next social influence factor can be seen to exert pressure, social proof or the tendency to conform to peer pressure.

Conformity

Cialdini's principle of Conformity or Social Proof refers to the tendency for people to do what everyone else is doing, to be persuaded or influenced by the prevailing consensus; this is group or peer pressure. The power of conformity to group pressures was demonstrated in a series of experiments conducted in the 1950's by another American psychologist, Solomon Asch.[17] In his studies, subjects were asked in a group setting to state which of three lines on a diagram were the same as another one. The answer was always clearly one of three options: A, B or C. Different items were presented in a

---

12 Sosabowski, *Freely I Served*, 139-143.
13 Baynes, *Urquhart of Arnhem*, 12.
14 Ibid., 32-33.
15 Urquhart, *Arnhem*, 16.
16 Buckingham, *Arnhem 1944*, 31-35.
17 Hewstone et al, *An Introduction to Social Psychology*, 247-249.

series of rounds. At first, all seven of the participants in the study (including the subject and six confederates of Asch's) gave the correct answers, then the six confederates began to give incorrect answers. The point of the experiment was to examine how the real subject would react when the confederates gave the incorrect answers. For example, the correct answer in any given round might be 'B'. The subject had written down 'B', but the confederates then said 'A'. What did the subject say when it was his or her turn, especially when the answer was clearly 'B'? In these experiments, across the 12 rounds where the confederates gave an incorrect response, 33 per cent of subjects conformed on each occasion (and said 'A'); and 75 per cent did so at least once. As with Milgram, these findings have been repeated across different conditions; conformity to group pressure is therefore a powerful mechanism, one that may have influenced Urquhart's actions and his intention to undertake the mission. So, what was the general atmosphere within Urquhart's command?

1st Airborne's Frustration

Within 1st Airborne Division, there had been some dissent, but this had been voiced about Operation Comet, which as we have seen, was the last of 16 operations involving 1st Airborne cancelled since June 1944, much to the division and Urquhart's frustration. Comet had immediately preceded Market Garden and involved the same basic operational concept and objectives. The key difference was that it had not involved the two American airborne divisions, employing only 1st Airborne Division and the Polish brigade to seize all the bridges along the corridor. The plan had also included dropping at night and *coup de main* assaults by glider.[18] There had been, however, issues and problems with the plan, that would be repeated for Market Garden. The landing and drop zones were several miles from their objectives, the bridges, with little consideration seemingly given to German reaction.[19] Overall, the operational area had been too big for one division and had invited failure.[20] The plan had called for 100 per cent mission success for each element, which experience had shown would have been highly unlikely. Ultimately, the plan was high risk; to work it needed virtually no resistance and ignored lessons learned from previous experience.[21] Concerns had been raised.

---

18 Harclerode, *Arnhem*, 26.
19 Buckingham, *Arnhem 1944*, 69.
20 Bennett, *A Magnificent Disaster*, 20.
21 Ritchie, *Arnhem*, 108-112.

Major-General Sosabowski, commander of 1st Polish Independent Brigade and Brigadier Shan Hackett, commander of 4th Parachute Brigade (part of Urquhart's division) had both voiced strong concerns about the plan. They had been concerned about its underestimation of the enemy and lack of consideration to any response to the proposed landings. Both officers had wanted more troops landed.[22] Frustrated, Sosabowski had interrupted one of Urquhart's briefings, exclaiming: 'But the Germans, General ... the Germans!'[23] Sosabowski had been so concerned that he had even confronted Urquhart and asked for his orders in writing, saying: 'General, I received your orders altering the original plan and I would like you to confirm these orders in writing'.[24] Hackett had been present at this exchange and agreed with Sosabowski.

Urquhart's Concerns

Urquhart was also suitably concerned. He took Sosabowski to see Browning who although he agreed, stated he had no more troops available and fobbed them off with the glib Red Devils comment.[25] The quotes from Sosabowski's memoirs again need to be treated with caution as he came in for criticism after the operation, was badly treated by the British establishment and became very embittered.[26] His account though, aligns with the memoirs of others, including Urquhart.[27]

Hackett was also unhappy, later commenting: 'The airborne movement was very naïve' and that they were 'innocents when it came to fighting the Germans'. Perhaps most damningly of all, he would also assert that: 'They used to make a beautiful airborne plan and then add the fighting-the-Germans bit afterwards.'[28] There was little enthusiasm for the operation in other quarters also. At one briefing, a company commander was reported to have whispered to another, 'That should provide you with either a Victoria Cross or a wooden one'.[29] Concerns about the viability of Operation Comet had been clearly raised. Doubts had begun to emerge when Second

---

22 Bennett, *A Magnificent Disaster*, 20.
23 Powell, *The Devil's Birthday*, 41.
24 Sosabowski, *Freely I Served*, 142.
25 Ibid., 143.
26 Middlebrook, *Arnhem 1944*, 447-448.
27 Urquhart, *Arnhem*, 17.
28 Middlebrook, *Arnhem 1944*, 8.
29 Powell, *The Devil's Birthday*, 41.

Army renewed its advance on 6th September, when it became clear that German resistance had stiffened.[30] Comet was duly cancelled on 10th September, replaced by Operation Market Garden. The situation and the atmosphere had now changed.

At the briefing for Market Garden delivered by Urquhart on 12th September, crucially, the tone was different. There was a lack of dissent. Sosabowski recalls in his autobiography, 'At the end Urquhart asked: "Any questions?" Not one brigadier or unit commander spoke. I looked round, but most of them sat nonchalantly with legs crossed, looking rather bored and waiting for the conference to end'.[31] Urquhart himself states that by the end of the Orders Group: 'Generally, the brigadiers appeared not unhappy with the task confronting us and they now dispersed to make their own brigade plans'.[32] As Sosabowski points out, no one in the airborne community was now seriously challenging the air plan.[33]

We would suggest that this incident is a good example of Social Proof playing a role in exerting pressure on Urquhart to go ahead with the operation. To Sosabowski, Urquhart was now proceeding with a sense of confidence that he described as foolhardy.[34] When challenged, Urquhart justified his confidence by citing intelligence reports that there were no German forces in the area.[35]

In summary, whereas concerns had been raised about Comet, the situation had now changed. There were, therefore, no stridently dissenting voices whilst Urquhart was planning his division's role in Market Garden. Acceptance of risk is crucial in any military operation, especially in airborne operations. As was demonstrated with the Solomon Asch experiments on conformity and social pressure, however, the silence that surrounded him was likely to have applied more pressure on Urquhart to carry through the allotted task. He would have lacked social support for raising any concerns and lacked any encouraging guidance from other dissenting voices. The string of prior cancelled operations had served to create a dangerous sense of momentum, one that was very much in line with the next influence factor we shall consider, Commitment and Consistency.

---

30 Clark, *Arnhem*, 24.
31 Sosabowski, *Freely I Served*, 145.
32 Urquhart, *Arnhem*, 11.
33 Sosabowski, *Freely I Served*, 196.
34 Hibbert, *The Battle of Arnhem*, 44.
35 Buckingham, *Arnhem 1944*, 69.

Commitment and Consistency

Another factor that would have applied pressure to Urquhart relates to Cialdini's principle of Commitment and Consistency. As we have seen with the discussion on Browning, this refers to the tendency to carry on with an action when you have already committed some time and energy to it; you typically feel the need to be consistent with your prior commitments. In this instance, the impact of the 16 previously cancelled operations can be seen to be a factor for Urquhart. It is worth quoting in full from his memoirs: 'By September 1944 my division was battle-hungry to a degree which only those who have commanded large forces of trained soldiers can fully comprehend [...] We were ready for anything. If there was a tendency to take light-heartedly the less encouraging factors, and even the unknown ones, it was understandable. Certainly, it is impossible to over-emphasise the ultimate significance of this procession of operations that never were'. He goes on to add that: 'Only the participant can adequately apportion the invisible factors, such as the effect of the sixteen cancelled operations in a row.'[36] These quotes neatly encapsulate Urquhart's need to see all that waiting, disappointment and hard work come to fruition, and to go ahead with the operation. To add to this, as we have seen, the operational window of opportunity was closing, which takes us on to the last of Cialdini's principle we shall explore – Scarcity.

Scarcity

Cialdini's principle of Scarcity states that a perceived time pressure or lack of a commodity will generate a demand for it. The idea is that individuals are influenced to act so that they do not miss out on the object that is of interest to them.[37] As with the principle of Commitment and Consistency, as a member of FAAA, it is likely that Urquhart and 1st Airborne Division would have been subjected to this influence factor.

Closing Window

By September 1944, the opportunity for 1st Airborne Division to see action, particularly in the role it had trained for, was rapidly diminishing. After the breakout in Normandy and the pursuit of the Wehrmacht across northwest

---

36 Urquhart, *Arnhem*, 18.
37 Cialdini, *Influence*, 203-231.

Europe, the war looked as if it was ending. There was, therefore, a real chance that the war might be over before his division could make an operational drop. Adding to this anxiety was the fact autumn was rapidly approaching and the weather in northwest Europe would soon be deteriorating, making airborne operations impossible. If 1st Airborne Division didn't go soon, they would have to wait for maybe six months before the weather improved with spring; with the received wisdom regards the state of German resistance, it seemed likely that the war would be over by then. Another and perhaps more insidious concern was added to his worry about the likelihood of the war ending. The British manpower situation had now become so acute, that 1st Airborne Division risked being employed as ordinary line infantry or being disbanded and used to reinforce other units. Against this backdrop of manpower problems, 1st Airborne Division looked like a very tempting pool of reinforcements.

Threat of Disbandment

Michael Carver, who took part in Market Garden as an armoured brigade commander in Second Army, stated that, in terms of 1st Airborne, in his opinion: 'If they were not employed now, they foresaw disbandment and the use of their troops as normal infantry if the war went on much longer'.[38] This view is further supported by an article in the British Army Review: 'Had 1st Airborne Division not been committed to Market Garden (or something like it) pressure to bring it into battle as an infantry division, or worse, to break it up, would have become acute'.[39] The imminent threat of being put into the line as normal infantry or indeed disbanded, would have served to further shorten the window of opportunity that was open to 1st Airborne Division. This, combined with the belief that German resistance was nearing an end, and autumn was approaching, would have created a keen sense of urgency within the division (comprised primarily of men who had volunteered to see action) and undoubtedly contributed to its eagerness to take part in an airborne operation soon. Urquhart was part of this.

---

[38] Michael Carver, *The Seven Ages of the British Army* (London: Harper Collins, 1986), 277.
[39] "Carbuncle", 'On an Excess of Bridges', *British Army Review*, No.108, 89.

Conclusion

Cialdini's model of social influence factors has been a useful framework to explore the situational pressures that Urquhart experienced when addressing the problem he faced. The orders he received meant he was required to be obedient to the explicit authority of the military command structure. The interest in the operation at senior levels meant that he was subject to a good deal of implicit authority as well. Added to this, the lack of dissension about the plan meant that there would have been a lot of pressure for him to conform to the prevailing mood of optimism. The formation of the Airborne Forces in general and FAAA in particular meant that there was a strong commitment to using 1st Airborne in the role for which it had trained; this pressure was only increased by the 16 operations that had been cancelled for the division during the summer of 1944. Finally, more pressure was added due to the window of opportunity to use the division in the airborne role was closing as the war appeared to be ending soon and the division was threatened with disbandment. Urquhart therefore faced a lot of pressure to proceed with the operation. He was not unique in facing these pressures and it could be argued that regardless of the situation, he should have been more robust in fighting his corner. To understand why he did not, it is important to understand his personal characteristics, the 'filters' that operate during the second, Orientation stage of the OODA Loop. These will be examined in the next chapter.

12

Urquhart's Orientation – Complexity

'Infantry rules operated, if under different conditions'.[1]

Introduction

Urquhart's Headquarters, Grantham, England, 12th September 1944

Roy Urquhart had just completed his Orders Group, outlining his plan for his formation's role in Operation Market Garden to the senior officers in his division. They were a tough audience, many of them were experienced airborne officers and he was not, but they seemed relatively happy with the plan that he had outlined. Putting the plan together had been a tough challenge, made worse by only having a few days to work with. The operation was massive in scale, bigger than the airborne component of the D-Day airborne operation and was suitably complex because of this. Given the prize, Arnhem, he was under a lot of pressure to succeed.

Capacity

To understand how Urquhart reacted to the pressure he was under, it is important to understand the personal characteristics he would have used to make sense of the situation. Urquhart has been described as out of his depth in command of 1st Airborne. This chapter, therefore, explores the way one of Urquhart's personal factors affected the way he interpreted the situation and influenced the conclusions he reached, his cognitive capacity. In this chapter we will examine the extent to which his underlying ability to think at the right

---

1   Urquhart, *Arnhem*, 15.

level of complexity impacted on his planning; whether his problem-solving matched the demands of the stressful situation he faced.

Task Complexity

Different models of decision-making in stressful situations (such as military battles) are based on the notion that problems faced by individuals in such scenarios fall along a continuum of complexity.[2] At the lower end of the continuum, the problems are simpler, more concrete in nature and operate on a smaller scale. The situation is more tactical and requires a more direct approach to command, with a fast, more heuristics-based thinking style where previous experience is helpful to generate solutions. This type of problem-solving has been referred to as System One-type thinking.[3] At the other end of the continuum, the situations are more complex, ambiguous and larger in scale.[4] Because they are complex, these situations require a more deliberate and analytical approach, which is referred to as System Two-type thinking.[5] Different situations, therefore, vary in terms of the complexity of the demands they place on an individual, as well as the complexity of the information-processing required to resolve the situation.[6]

Stratified Systems

The concept of a continuum of task complexity is central to Elliot Jaques' Stratified Systems Theory. This suggests that within an organisation, information-processing and decision-making takes place along a continuum of complexity ranging from concrete to the more abstract. Jaques divided this continuum into specific strata with each level involving increasing amounts of complexity and different types of thinking.[7] At the bottom of the continuum, Level One, thinking is based on simple rules and anchored in tangible output. At the next level, a decision-maker needs to adapt existing rules to meet a given objective. At Level Three, you need to be able to extrapolate from given

---

[2] Lars Fredholm, 'Decision making in firefighting and rescue operations' in *Sitting in the Hot Seat*, ed. Rhona Flin, (London: Wiley, 1996), 158.
[3] Kahneman, *Thinking, Fast and Slow*, 19-30.
[4] Berndt Brehmer, 'Dynamic Decision-Making: Human Control of Complex Systems', *Acta Psychologica*, 81, (1992), 211-241.
[5] Kahneman, *Thinking, Fast and Slow*, 19-30.
[6] Kenneth Hammond, 'Judgement and decision making in dynamic tasks', *Information & Decision Technologies)*, 14, (1988), 3-14.
[7] Elliott Jaques, *Requisite Organisation* (Arlington: Cason-Hall, 1998), 72.

rules and create new connections within a defined system. At the next level up, you need to be able to develop alternative approaches and to evaluate their success against knowledge and experience. Level Five involves making relationships between previously unrelated sets of material. Finally, thinking at Level Six is based outside of the present and concrete problems; new ideas and realities are created.[8]

Jaques went on to link these different levels of task complexity with seniority within an organisation and, of interest to us, to ranks and command positions within the military hierarchy. According to Jaques, command of a section involves Level One thinking, a company, Level Two and a battalion, Level Three. Level Four thinking is required at brigade level, Level Five at divisional level and Level Six at Corps.[9]

Of interest to our discussion here is that Urquhart, as a divisional commander needed to be operating, in complexity terms, at Level Five. Upon taking up command of the division, his thinking needed to move away from utilising previous experience. We shall return to this issue later but think about the quote with which we opened the chapter. Urquhart needed to match the complexity of his thinking with the complexity of the new, more complex task of running a division. Jaques calls this Requisite Complexity.[10]

Requisite Complexity

To work effectively within an organisation, you need to match the complexity of your thinking to the complexity of the task demands of your role. This process can be seen in a research study that examined the longer-term fortunes of 19 leaders in successful national revolutions. Each of the 19 leaders showed lower levels of complexity in the revolutionary phase where there was a need to shape a simple narrative for people to rally around. When the revolutionary leader was in government and now running the country, a more complex approach was required. Those that were successful at this stage were able to increase the complexity of their problem-solving; those that didn't, failed.[11] As some leaders were able to adjust and increase the complexity of their thinking whilst others were not, this study suggests

---

8 Ibid., 136.
9 Ibid., 136.
10 Ibid., 28.
11 Peter Suedfeld & David Rank, 'Revolutionary leaders: Long-term success as a function of changes in conceptual complexity', *Journal of Personality & Social Psychology*, 34, 2, (1976), 169-178.

that people vary in their ability to think in increasingly complex terms; this capacity is referred to as trait or Conceptual Complexity.

Trait complexity

Conceptual Complexity is defined as a psychological trait, a stable psychological characteristic which determines the maximum amount of complexity that you can bring to bear in any given situation. [12] Within this latent capacity, the level of complexity that you deploy to solve a problem is your state level of complexity.[13]

State complexity

The actual level of complexity that you apply to a problem, the 'state' level of complexity, is called Integrative Complexity.[14] High levels of state or Integrative Complexity are enabled by high levels of Conceptual Complexity; if you have a capacity for thinking in complex terms, you can deploy this capacity to solve complex problems. On the other hand, if you have a low level of Conceptual Complexity, you don't have much capacity to think in complex terms, then you would only be capable of operating at a low level of Integrative Complexity.[15]

Integrative Complexity is defined as: 'the level of complexity shown in thought and behaviour in any particular context or situation'.[16] Integrative Complexity is seen as a dimension or a sliding scale of complexity. At the lower end of the dimension, the thinking is simplistic. It is characterised by rigidity in thinking, making of gross distinctions, restricted information usage and simple responses. There is more complex thinking at the higher end

---

[12] Christopher Porter & Peter Suedfeld, 'Integrative Complexity in the Correspondence of Literary Figures: Effects of Personal and Societal Stress', *Journal of Personality & Social Psychology*, 40, (1981), 321-330.
[13] Peter Suedfeld & Philip Tetlock, 'Integrative Complexity of Communications in International Crises', *Journal of Conflict Resolution*, 21, 1, (1977), 169-184.
[14] Peter Suedfeld & Philip Tetlock, 'Integrative Complexity of Communications in International Crises', 169-184.
[15] Peter Suedfeld, 'Cognitive Managers and Their Critics', *Political Psychology*, 13, 3, (1992), 435-451.
[16] Peter Suedfeld et al, 'Conceptual/integrative complexity' in *Motivation and Personality: Handbook of thematic content analysis*, ed. Charles Smith, (New York: Cambridge University Press, 1992), 393-400.

of the dimension. At this end, you have much greater flexibility in thought, make more nuanced distinctions, provide more complicated responses and engage in a more extensive information search and usage.[17]

Integrative complexity progresses through seven distinct levels. The lowest level means there is no differentiation of the issue into different dimensions or perspectives. At this level you will believe that there is only one reasonable approach to the problem, expressing your view in absolute or categorical terms. There is little or no acceptance of ambiguity. A step change occurs when you show moderate or even high differentiation, this is level three. Here, you can identify at least two alternative perspectives or different dimensions. You begin to see connections between the different dimensions at level five. Now, you can view alternative factors interactively and recognise the need to consider mutual interactions and causal relationships. At level seven you adopt a global overview of the situation, can think laterally and make complex trade-offs between different factors. The three other levels that make up the dimension are seen as transitional stages between the main levels where ideas are implicit and emerging and not explicit or well defined.[18]

Higher levels of Integrative Complexity can be compared to the Vigilant coping pattern in the Decisional Conflict Model (we touched on this briefly in Chapter Three). According to the DCM, Vigilant thinking involves decision-making that meets seven criteria for effective information processing, which involve the generation and comparison of multiple alternatives (and thus higher levels of Integrative Complexity). Failure to meet one of these criteria constitutes a defect in decision making and the more criteria missed the more likely is it that the decision outcome will be ineffective. Janis and Mann suggest that each coping pattern in the DCM meets more or less of the seven information processing criteria.[19] The Coping Pattern – Vigilance is the only pattern that involves the generation and comparison of different alternatives or dimensions in a situation and so requires a high degree of Integrative Complexity.

---

[17] Theodore Raphael, 'Integrative complexity theory and forecasting international crises', *Journal of Conflict Resolution*, 26, 3, (1982), 423-450.
[18] Suedfeld et al, 'Conceptual/integrative complexity', 393-400.
[19] Janis & Mann, *'Emergency Decision Making'*, 35-48.

Research

Integrative Complexity research has addressed several different areas that are of relevance to this discussion. High levels of Integrative Complexity have been linked to international crisis situations that resulted in successful outcomes, whereas poor outcomes were linked to low levels. For example, the diplomatic communications made before the outbreak of the First World War in 1914 and the Korean War in 1950 were compared with those during the Moroccan Crisis in 1911, the Berlin Blockade in 1948, and the Cuban Missile Crisis in 1962. The communications produced during the first two crises were significantly lower in complexity than in the latter three.[20]

The Confederate General Robert E. Lee was able to defeat the Union forces set against him in the six battles where he demonstrated higher levels of Integrative Complexity than his opponent. When he showed lower levels than his opponents, he was defeated.[21]

Another study examined nine international crisis situations that led to one country mounting a surprise attack. The results indicated that the Integrative Complexity of communications made by the aggressor showed significant reductions between three months and one week before the attack.[22] Another study investigated the relationship between the United States and the Soviet Union over the issue of Berlin between 1946 and 1962, Integrative Complexity was found to decline just prior to the onset of the crises of 1948 and 1961 (construction of the Berlin Wall); after this, it increased again leading to a peaceful resolution.[23] This research suggests that poor outcomes are linked to the adoption of low levels of Integrative Complexity (and less Vigilant information-processing) by those in higher levels of organisations. One interesting question that we need to address is whether people are born with this capacity for high levels of complexity?

---

20 Peter Suedfeld & Philip Tetlock, 'Integrative Complexity of Communications in International Crises', 169-184.
21 Peter Suedfeld, Raymond Corteen & Carroll McCormick, 'The role of integrative complexity in military and leadership: Robert E Lee his opponents', *Journal of Applied Social Psychology*, 16, 6, (1986), 498-507.
22 Peter Suedfeld & Susan Bluck, 'Changes in integrative complexity prior to surprise attacks', *Journal of Conflict Resolution*, 32, 4, (1988), 626-635.
23 Raphael, 'Integrative complexity theory and forecasting international crises', 423-450.

Mode	Level	Command	Rank
Seven	Seven	Army	General
Six	Six	Corps	Lieutenant-General
Five	Five	Division	Major-General
Four	Four	Brigade	Brigadier
Three	Three	Battalion	Lieutenant-Colonel
Two	Two	Company	Major
One	One	Section	Sergeant

Figure 12.1   Modes of Complexity

Mode

We have discussed how the psychological trait of Conceptual Complexity defines the upper limit of Integrative Complexity that you can use in any given situation. Jaques suggested that this capacity increases as people mature, but there is a limit to the level to which a person can attain. Jaques called this developmental path the person's 'Mode'.[24] This notion is supported by other research which suggests that complexity continues to develop with age.[25] The title of the Mode is defined by the highest level that a person can effectively attain. These modes, as outlined by Jaques are shown in Figure 12.1.

For example, if you are a Mode Three you will only be able to progress to Level Three in the organisational task complexity hierarchy (or command of a battalion). You will be out of your depth if you are promoted beyond this level, such as a brigade (Level Four) or a division (Level Five). In this situation, you will be unable to cope effectively as your problem-solving will be too simplistic. The further you are promoted past your level of complexity, the worse your problem-solving and performance becomes. A good example of this is Sir Redvers Buller, who ineffectually commanded the British forces in the Boer War. He has been described as: 'a superb Major, a mediocre Colonel and an abysmal General'.[26] In Jaques' terms, he was a Mode Two

---

24 Jaques, *Requisite Organisation*, 29.
25 Porter & Suedfeld, 'Integrative Complexity in the Correspondence of Literary Figures: Effects of Personal and Societal Stress', 321-330.
26 Dixon, *On the Psychology of Military Incompetence*, 220.

and so effective in the rank of Major in the middle years of his career. As he progressed to higher commands, the gap between his complexity of problem-solving grew larger and he became increasingly more ineffective. So, what about Urquhart?

Urquhart's trait complexity

As the discussion above highlights, the maximum level of complexity that a person can bring to bear when problem-solving or engaged in information-processing (Integrative Complexity) is determined by his or her underlying capacity for thinking in complex terms (Conceptual Complexity), a capacity which increases with age (Mode). In terms of Urquhart, a high capacity for complexity would have allowed him to develop a comprehensive and nuanced understanding of the situation he faced. Alternatively, if he was too simplistic in his approach, he might not have fully appreciated the complexities and inter-dependencies of the problems with which he was confronted. His ability to appreciate and deal with the task at the correct level of complexity is therefore key to understanding how he oriented to the situation, and to answering the question of whether he was out of his depth. Before examining Urquhart's level of Integrative Complexity when he was planning 1st Airborne's role in Market Garden, we must first examine his Conceptual Complexity and Mode. There is evidence to suggest that Urquhart did not possess high levels of Conceptual Complexity or in Jaques' terms, a very high Mode. Urquhart was described by Miles Dempsey as the most vocal but not the most able of divisional commanders.[27] To examine this more closely, and to try to establish his Mode, we need to look at his career history.

Urquhart was unremarkable at school (St Paul's); he failed the entrance exam for the more specialist Royal Military Academy at Woolwich but passed the one for Sandhurst, where he made reasonable progress in academic subjects. He then enjoyed a steady career in the pre-war Army. He was lucky, due to circumstances, to be promoted to Captain after 8 years at the age of 27. Following on from this, new regulations meant that he was able to promote to Major at the age of 38 after 15 years of service (the ages are important). He was subsequently spotted by Montgomery and quickly promoted to Lieutenant-Colonel at the age of 39.[28]

---

27 TNA, WO 285/29, Dempsey to Ellis, 7 July 1966.
28 Baynes, *Urquhart of Arnhem*, 6-27.

Prior to his appointment to command 1st Airborne Division, he had only commanded a Brigade for six months. As an acting Brigadier, he had been appointed to command 231st Malta Brigade Group at the age of 42 in 1943.[29] As a Brigade commander in Sicily in 1943, Urquhart was constantly up front with his troops.[30] Urquhart would state in his memoirs that his command of the Malta Brigade was the most significant of his career. Clearly, caution must be exercised when utilising a point drawn from an individual's memoirs, but this does suggest that at this stage of his career, his level of Conceptual Complexity was in tune with the required level of task complexity. If this is the case, given his age and his previous history (Captain at 27 and Major at 38), this would suggest that in Jaques' terms, he was a Mode Four.

Urquhart was then appointed Chief of Staff of XII Corps back in Britain in late 1943. In this post, Urquhart got his subordinates to do most of the staff work; he states that he did not like staff work and was at his happiest when commanding troops on his own. By the time of Market Garden, he had commanded a division for eight months and an expanded one at that as he also had 1st Independent Polish Airborne Brigade under command.[31] According to Jaques' model, divisional command requires Level Five thinking which is characterised by the ability to make relationships between previously unrelated sets of material and, therefore, to create general theoretical rules which redefine fields of knowledge and experience. The key point here is a move away from previous knowledge and experience and the generation of different possibilities and novel solutions. In Jaques' model, command of a brigade involves Level Four thinking: the ability to develop alternative approaches and to evaluate their success against knowledge and experience.[32]

To be capable of operating at Level Five at his age (he was 42 in 1944), Urquhart would have needed to have possessed a high Mode (Mode Seven). Urquhart's early history and competence at brigade command level (in his early forties) suggest he did not have this capacity. Indeed, we have already suggested that he was in fact, a Mode Four. It is interesting to note that Urquhart had 'served with distinction in Sicily and Italy as a brigade commander' and had only just been promoted (January 1944) to Major-General and appointed to command 1st Airborne Division.[33] In terms of

---

29 Ibid., 48.
30 Ibid., 57-62.
31 Ibid., 49-66.
32 Jaques, *Requisite Organisation*, 136.
33 Buckingham, *Arnhem 1944*, 28.

seniority of rank, as Urquhart himself notes, when he took over command of the division, he was only an acting Major-General.[34] He had been, as a Lieutenant-Colonel, a battalion commander only two years before. This suggests that Urquhart would have been more comfortable operating at Level Four or below.

Further examining his past experiences suggest that he tended to look at problems and situations in a practical, realistic and factual manner (lower end of the complexity spectrum). A good example of this is the description of Urquhart by Major Henniker, CRE (Commander Royal Engineers), 1st Airborne Division, who described him as: 'A good practical soldier, with both feet on the ground, and not too much airy-fairy nonsense with him'.[35]

Baynes refers to Urquhart's time as the Adjutant for 1st Battalion of the Highland Light Infantry in the 1930s. He describes how the commanding officer, Lieutenant-Colonel Alec Teller-Smollett was very hands off in his approach and left Urquhart to manage the daily routine of battalion life. According to Baynes: 'This suited Urquhart well'.[36] This suggests the sort of more grounded approach associated with lower Modes and thus lower levels of complex thinking.

These descriptors point to a man who was no-nonsense and pragmatic in approach with a firm and realistic grasp on the detail of a situation but now faced the task of commanding an airborne division. This was a more complex and political task and (in an airborne role) in a situation that would be highly dynamic which posed unique challenges (such as fighting with less heavy weapons and without other resources typically available to ground infantry units).[37] This suggests Urquhart, faced with a requirement to solve a novel and complex problem (Level Five within Jaques' model), was predisposed, because of his current capacity for complex thinking (Level Four) to apply his previous knowledge and experience to the situation, which as we have seen, was infantry-based.

Urquhart's information processing

Urquhart's lack of airborne experience made his appointment seem a little unusual. Sosabowski states in his memoirs that Urquhart was a strange

---

34 Urquhart, *Rising Eighty*, 102.
35 Mark Henniker, *An Image of War* (London: Leo Cooper, 1987), 163.
36 Baynes, *Urquhart of Arnhem*, 13.
37 Buckingham, *Arnhem 1944*, 35-36.

choice, especially when there were experienced airborne brigadiers within the division.[38] Urquhart was obviously aware of his lack of airborne experience, stating in his memoirs: 'I was very much aware of my lack of experience of airborne operations'.[39] He raised his concern over his suitability with Browning, who dismissed them.[40] Lloyd Clark suggests that this lack of experience meant he could be manipulated by Browning.[41] It also had a direct impact on his planning for Market Garden.[42] Dempsey goes so far as to suggest that the failure at Arnhem was due in large part to 'inept' planning by Urquhart and his staff.[43] His lack of experience meant that he found it difficult to judge the feasibility of proposed operations or to impose his authority on his staff and subordinates.[44] As we have seen, he did have staff and brigade group command experience, but this was infantry and not airborne.

There were certainly misgivings about Urquhart in 1st Airborne, Lieutenant-Colonel Frost commented that Urquhart did not truly grasp the airborne concept. This, at its most fundamental meant an airborne formation would need to operate on its own for 48 hours without artillery support, a task an infantry unit is not generally required to undertake.[45] This is the key issue. When faced with a Level Five problem, Urquhart, a Mode Four, applied Level Four thinking and relied on his past infantry experience. The problem was the two situations were different; as Lloyd Clark points out: 'Urquhart held some damaging and unchecked assumptions' about the similarities.[46] As Urquhart himself states in his memoirs: 'It seemed to me that the basic rules were the same'. Viewing an airborne division as highly trained infantry, his view was that 'infantry rules operated, if under different conditions'.[47] These statements neatly describe the sort of Level Four thinking that involves the development of 'alternative approaches and to evaluate their success against knowledge and experience'. The problem was that the analogy did not hold, the two situations (conventional ground warfare and airborne operations) were different.[48] As John Frost explains: 'The snag of bringing in a complete

---

38 Sosabowski, *Freely I Served*, 196.
39 Urquhart, *Arnhem*, 15.
40 Buckingham, *Arnhem 1944*, 52.
41 Clark, *Arnhem*, 115.
42 Buckingham, *Arnhem 1944*, 51.
43 TNA, WO 285/29, Dempsey to Ellis, 18 June 1962.
44 Buckingham, *Arnhem 1944*, 34.
45 Baynes, *Urquhart of Arnhem*, 73-82.
46 Lloyd Clark, *Arnhem*, 115.
47 Urquhart, *Arnhem*, 15.
48 Buckingham, *Arnhem 1944*, 35.

newcomer was that however good they might be, they were inclined to think that airborne was just another way of going into battle'.[49] He goes on to point out that an airborne unit is separated from the usual support functions, specifically the supply of ammunition and medical care.

One key outcome from all of this was that Urquhart's lack of airborne experience also meant that he was less able to query and challenge the restrictions placed upon him by the air plan. As Lloyd Clark points out he could be: 'manipulated far more easily by I British Airborne Corps than those with an airborne pedigree. The downside, however, was that he also offered less constructive criticism'.[50] The ramifications of this mismatch will be explored next in terms of the problems that manifested themselves in Urquhart's plan.

Conclusion

In summary, Urquhart's level of cognitive capacity ahead of Market Garden was misaligned with the task complexity he faced. His low to moderate Conceptual Complexity or Mode Four thinking (a tendency to apply previous knowledge and experience to problems) meant his thought processes were non-optimal now he faced a novel and more complex (Level Five) situation. Urquhart was predisposed to look at the problems he faced in a more simplistic and rigid manner and rely on experience that was not wholly relevant, his previous (infantry, not airborne) background probably led him to make inappropriate sense of the issues he needed to address.

Having made sense, in his terms, of the challenges he faced, he then had to cope with them. This brings us to the third stage of the OODA Loop – Decide; we will explore Urquhart's coping strategy for this stage in the next chapter.

---

49 Frost, *A Drop Too Many*, 195.
50 Lloyd Clark, *Arnhem*, 115.

13

Urquhart's Decision – Hypervigilance

'The plan had to be produced quickly'.[1]

Introduction

Edwin's Tale

This is how the story goes. The date, 15th September 1944, two days before Market Garden. The place, General Browning's office in his headquarters at Moor Park golf course in northwest London. Roy Urquhart, marched into Browning's office and told his boss: 'Sir, you have ordered me to plan for this operation and I have done so to the best of my ability. But now, as far as I can see, it is a suicide mission'. With that, he promptly turned round and marched out.[2] The scene was witnessed by Captain Edwin Newbury, Browning's ADC. If true, it is quite an extraordinary tale and one that provides a fascinating window into Urquhart's state of mind just 48 hours before he embarked upon Operation Market Garden. Here was a professional soldier telling his commander that he was being sent on a suicide mission. Five days before, Urquhart had begun planning his division's role in the operation with an optimistic frame of mind. He was now, with only two days to go, very pessimistic, if not fatalistic in his appraisal of the situation. How did this change of mind come about?

This chapter discusses Urquhart's thinking in the third Decide stage of the OODA Loop. We will again use Janis and Mann's Decisional Conflict Model as the framework through which to explore the coping strategy he

---

[1] Urquhart, *Arnhem*, 9.
[2] CRCP 108/5, Cornelius Ryan Collection of World War II Papers, Mahn Center for Archives and Special Collections, Ohio University, Athens, Ohio.

used to deal with the pressures he was under in planning his division's part in the operation. The first question, then, that we need to address is Urquhart's perception of the level of risk that 1st Airborne Division would face at Arnhem.

Risk

Urquhart's Risk Calculus

In terms of Urquhart's risk calculus, a couple of key questions come into sharp focus. What was Urquhart's perception of the risk his division faced at Arnhem and perhaps more importantly, what picture was being painted to him by his higher formation? The answers to these questions, therefore, hinge on the intelligence picture for Market Garden. The issues around the quality of the intelligence available to Urquhart and 1st Airborne for Market Garden and its use are complex and remain topics of heated debate, very prevalent in the Arnhem literature. It presents to some extent, a contradictory picture.[3] For example, Urquhart's assessment of the risks facing his division was made against the general backdrop of victory euphoria and the optimistic intelligence assessments being made at the time, such as the 'end of the war is in sight' SHAEF intelligence summary from 26th August.[4] On the other hand, the cancellation of Comet due to stiffening resistance was a key point that brought a different perspective to Urquhart's perception of risk and hence his adoption of a DCM coping strategy. As we have seen in our discussion on Browning, there was some recognition, at least at senior command levels, that the German forces in the area possessed renewed fighting capability and therefore posed a risk to the operation. Staff at I Airborne Corps were aware of the armoured strength at Arnhem even though the Ultra decrypts were not passed down to them; the Dutch underground reports were dismissed; and the aerial reconnaissance photographs were questioned or downplayed. As we have seen, the intelligence summary appended to the order for the operation referenced the presence of the 50-100 Mark IV's in the Arnhem area.[5] Browning, was informed of the presence of II SS Panzer Corps in the area on 10th September; and then crucially that 9th SS Division was at the equivalent strength of a brigade group but with few tanks. On

---

3   Dickson, '"But the Germans, General, the Germans, what about them?"', 116.
4   TNA, WO 219/1922, SHAEF Intelligence Summary No.23, 26th August 1944.
5   TNA, WO 106/972, 1st Airborne Corps Operation Instruction No.1 Para 1.

13th September, I Airborne Corps passed this information onto Urquhart (crucially lacking the formation designations),[6] whilst also stating German troops would be incapable of any 'organised resistance'.[7] Staff at I Airborne Corps thus passed on an accurate assessment of the size of the German forces at Arnhem, but failed to underscore the quality of the units, indeed the fact that they were Waffen SS troops was not mentioned.[8]

The risks were known at higher formation level (for example by Brian Urquhart and Browning at Corps and Bill Williams at Army Group level) and there was within the Allied (airborne) community an acceptance that German forces in the area posed a risk to the operation, but the risk was played down. The reason was expediency, and the expediency was due to the eagerness of 1st Airborne to get into action driven by frustration from the previously cancelled operations.[9] As Urquhart himself states in his memoirs, his division was 'battle-hungry'.[10]

First Airborne were being called the 'Stillborn Division',[11] and as Frost states, it was delighted with Market Garden, viewing it as a 'daring airborne operation'.[12] Even those previously critical of Comet were now, with its additional US airborne divisions, ready to go, as Brigadier Hackett related: 'We realised that we had to get into battle after all those cancellations ... [y]ou had to get into battle almost at any price'. This meant that, in his view 'shortcomings in the plan were readily forgiven as long as we could get in there'.[13] As stated, a threat of disbandment due to the manpower crisis, added a degree of urgency.[14]

The risk was also downplayed due to the high levels of confidence within 1st Airborne. Some historians have argued that members of the division were over-confident, perhaps even cocky and arrogant.[15] Buckingham argues this confidence was based on wishful thinking.[16] Henniker, who commanded an engineer component in the Mediterranean felt the division 'surrounded

---

6 TNA, WO 171/133, 21 Army Group Intelligence Review, No.160, 18 Sep 1944.
7 WO 171/393, 1 Airborne Division Report on Operation Market, Part V, HQ 1st Airborne Division War Diary, 1st-30th September 1944.
8 TNA, WO 171/393, 1 Airborne Division Planning Intelligence Summary No.1, HQ 1st Airborne Division War Diary, 1st-30th September 1944.
9 Baynes, *Urquhart of Arnhem*, 77.
10 Urquhart, *Arnhem*, 18.
11 Ryan, *A Bridge Too Far*, 121.
12 Frost, *A Drop Too Many*, 198.
13 Clark, *Arnhem*, 113.
14 Peaty, *Operation Market Garden*, 70.
15 Buckingham, *Arnhem 1944*, 48.
16 Ibid., 45.

themselves with a mystique that was not entirely justified by experience'.[17] Major Tower, recently posted into the unit agreed: 'We found the airborne boys, with their red berets, etc, hard to convince that other people had done a lot of fighting in the war. They were a marvellous lot, but they overestimated their prowess'.[18]

By September 1944, Urquhart knew his officers and men were desperate to get into action; he shared the same sentiment, commenting in his memoirs that 'we were ready for anything'.[19] He added: 'By the time we went on Market Garden we couldn't have cared less. I mean I really shouldn't admit that, but we really couldn't ... we became callous'. In terms of risk appreciation, this meant: 'We had approached the state of mind when we weren't thinking as hard about the risks as we possibly had done earlier'.[20]

Urquhart's optimism

These statements clearly suggest a failure to engage in the sort of rigorous risk assessment that we have been discussing; the key point, here, is an emerging picture of an optimistic appreciation of the risk, at least initially. These two statements were made by Urquhart after the war; there is, obviously, a possibility that he was attempting to defend his actions in his memoirs, but these comments are more admissions of guilt rather than defensive comments. Other than Urquhart's memoirs, can we look further afield for other evidence that we can use to substantiate the picture?

Sosabowski's 1st Polish Independent Airborne Brigade was attached to 1st British Airborne and would drop later at Arnhem; his recollections add to this picture of optimism. His memoirs describe Urquhart's reference to a lack of German resistance in the Arnhem area and that overall 'the operation represented [...] a risk we can well afford to take'. Sosabowski states he was concerned over this optimism; upon raising his concerns about the size of the division's perimeter once it had landed, Urquhart agreed but reassured him there would be no heavy German resistance.[21] Sosabowski would later describe Urquhart's confidence as 'almost foolhardy';[22] Gavin would appear

---

17 Ibid. *1944*, 31.
18 Middlebrook, *Arnhem 1944*, 22.
19 Urquhart, *Arnhem*, 18.
20 Baynes, *Urquhart of Arnhem*, 100.
21 Sosabowski, *Freely I Served*, 146-147.
22 Hibbert, *The Battle of Arnhem*, 44.

to agree with this sentiment.²³ These comments, however, paint a somewhat misleading picture because it does appear Urquhart did have some concerns.

Urquhart's concerns

Urquhart seems to have been aware that he was not being told the full story about the opposition he was likely to face at Arnhem, stating in his memoirs, '[I]n the division there was a certain reserve about the optimistic reports coming through from Twenty-First Army Group'.²⁴ As with other evidence drawn from Urquhart's memoirs, this statement should be viewed in the context of a possible attempt at post hoc justification, it does however, match the overall intelligence picture that had been passed to 1st Airborne Division – that reasonably strong, if depleted, German (armoured) units were near the Arnhem area. There is evidence suggesting that Urquhart and his brigadiers were told that there were two depleted Panzer Divisions in the area.²⁵ The 1st Airborne Division Planning Intelligence Summary 2 of 7th September mentions 50 tanks and a 'fair quota of Germans'.²⁶

A German After Action Report (AAR) further supports the idea that Urquhart and 1st Airborne Division had some appreciation of the German threat in the area. This report refers to intelligence found on captured 1st Airborne Division officers that referred in detail to German battlegroups in the Arnhem and Nijmegen areas;²⁷ information derived from a Second Army Intelligence Summary.²⁸ Urquhart himself provides further evidence he knew the German opposition was likely to be stronger than was being suggested at higher levels, stating in his memoirs that he had 'no illusions about the Germans folding up at the first blow'.²⁹ Again, this could be a defensive comment, written after the event, but it does corroborate the points made above. His statement should also be considered within the context that historically, the Wehrmacht had always fought doggedly in defence, were now defending their own border and so were likely to put up a stiff resistance. Buckley states that commanders from battalion upwards were

---

23 Gavin, *On to Berlin*, 150.
24 Urquhart, *Arnhem*, 7-9.
25 Bennett, *A Magnificent Disaster*, 198.
26 TNA, WO 171/393, 1st Airborne Division War Diary, Planning Intelligence Summary 2 of 7 September 1944.
27 TNA, AIR 20/2333, 16th SS Panzer Grenadier & Reserve Battalion Report, Arnhem.
28 LHCMA, Dempsey Papers, British 2nd Army Intelligence Summary, No. 113, 25 September 1944, Part 2.
29 Urquhart, *Arnhem*, 9.

briefed about the presence of Germans in the area.[30] Indeed, it seems that there was enough recognition of the threat within the division that some officers acted accordingly. Brigadiers Lathbury and Hackett briefed their battalion commanders to expect 50 per cent casualties.[31] Captain Eric Mackay of the Royal Engineers, ordered his men to take double the normal ammunition load and briefed them on escape and evasion techniques.[32]

Let's look at the problem from Urquhart's perspective. In summary, Urquhart does seem to have been aware that German forces in the Arnhem area posed a risk to his division when he and his staff were planning the operation. Although he did not receive full details of the fighting ability of the troops concerned, he was given reasonably accurate information about their composition. If we look at the situation through the lens of the DCM model, it seems clear Urquhart was aware of a clear risk to his division's role in the operation and thus he did not adopt an Unconflicted Coping strategy (either Adherence or Change). The inherent risk involved in airborne operations was compounded by the presence of capable enemy forces in the area and limitations inherent in the air plan, (his division arriving in three lifts thus reducing its combat power), to create a very real threat that 1st Airborne's move on Arnhem would be interdicted and stopped. Yes – there was risk; could he come up with a plan to mitigate it? Feasibility would be the next key issue.

Feasibility

Urquhart was facing a tricky problem, the air plan for Market Garden presented him with several challenges. A key issue was the shortage of available troop-carrying aircraft. He had the use of a total of 16 squadrons from 38 and 46 Group of the RAF for his glider-borne element and aircraft from IX US Troop Carrier Command for his paratroopers.[33] This was not enough. The shortage of aircraft was bad enough; the situation was further compounded by Major-General Paul Williams' (commanding IX US Troop Command) decision to fly one mission per day due to concerns over crew fatigue and the need for aircraft maintenance.[34] Urquhart had to deal with further constraints. The size of air lift available to 1st Airborne Division was also limited by the decision to give the two American airborne divisions a

---

30 Buckley, *Monty's Men*, 216.
31 Buckingham, *Arnhem 1944*, 93.
32 Ryan, *A Bridge Too Far*, 160.
33 TNA, WO 205/313 21st Army Group Operation 'Market Garden' Plans and Instructions.
34 TNA, WO 205/874, IX Troop Carrier Command Report on Operation Market.

greater allocation of aircraft. As Browning had pointed out to him, Market Garden was a 'bottom to top' operation, failure in the early stages would lead to disaster further up the ground corridor.[35] It was a decision Urquhart, quite naturally didn't agree with. The limited air lift capability meant his division could not be transported in one go.

Urquhart had to insert his division over three days. He would only be able to bring in two brigades on the first day. The best option as far as he could see was to bring in 1st Parachute Brigade and (most of) 1st Airlanding Brigade. With the rest of his command coming in over the next two days, the division would, therefore, need to hold the landing and drop zones until day three; Urquhart assigned this task to elements of 1st Airlanding Brigade. Crucially, this meant that combat power for the assault on the bridges in the vital first few hours, would be reduced to one brigade.[36] He knew this constraint would create severe difficulties for him if he faced tough opposition upon landing. But that wasn't the end of it.

There were more problems with the location of his landing and drop zones. The zones were a worryingly long way from the bridges. Again, there was not much he could do, the insertion plan had been dictated to him by the air planning staff who had several concerns. The flak batteries around the airfield at Deelen had been assessed as posing a threat to the aircraft. The urban built-up area to the north of the bridge at Arnhem, and the nature of the terrain to the south had been assessed as being unsuitable for parachute and glider landings. This meant that Urquhart was forced to select zones between six and eight miles away from his objectives – the road and rail bridges at Arnhem.[37] (One of the persistent myths about Market Garden is that the precise locations of 1st Airborne Divisions landing and drop zones were imposed on Urquhart. This is untrue, restrictions were placed on him, but as Urquhart later stated, he chose the locations).[38]

Urquhart was aware of the severe constraints these issues placed on his freedom of action. Urquhart would later reflect in his memoirs: 'I checked the distances to the target and found that this ground was between six and eight miles away – a formidable distance'.[39] This distance, a two hour approach

---

[35] TNA, WO 205/313, Operation Instruction No. 2 – British Airborne Corps, 14th September 1944.
[36] TNA, WO 171/393, 1st Airborne Div Op Instr No.9, HQ 1st Airborne Division War Diary, 1st-30th September 1944.
[37] TNA, WO 171/393, 1st Airborne Division War Diary, September-December 1944.
[38] Urquhart, *Arnhem*, 7.
[39] Ibid., 7.

march for heavily equipped men, meant Urquhart almost certainly risked losing the element of surprise, allowing German troops in the area (reacting quickly and decisively), sufficient time to respond to the landings and put blocking forces in place.[40] This would be a big problem.

The element of surprise would be further weakened by higher command's refusal to sanction a *coup de main* operation. Airborne doctrine and received wisdom at the time emphasised *coup de main* tactics for seizing bridges; it's what the Germans did in 1940 and the Allies in June 1944.[41] As had been explained to Urquhart several times, this restriction had been imposed on him for three reasons. Firstly, the timing of the drop meant that any *coup de main* party would have to be dropped in daylight making it vulnerable to ground fire. Secondly, concerns over the proximity of flak in the area (especially at the airfield at Deelen nearby) meant this option was ruled out as too dangerous. Thirdly, the nature of the terrain around the bridges would lead to concerning casualty rates (the urban nature of Arnhem to the north and the marshy Polder land to the south).[42] So, what did all this mean? The bridges, Urquhart concluded, would have to be seized by the three battalions of 1st Parachute Brigade advancing on foot from their drop zones between six and eight miles away.[43]

Positives

Criticism has been levelled at Williams and the planners at FAAA for generating a plan that failed to sufficiently factor in the needs of the airborne forces.[44] Sebastian Ritchie has challenged this, suggesting the plan was in fact the correct one given the circumstances and the problems facing the planners.[45] It probably doesn't matter whether this criticism is warranted or not; the key point is the air plan imposed severe constraints on Urquhart's planning ability and therefore brings into question the probability of him developing a viable solution. To mitigate against this pessimistic view, a combination of factors would have led Urquhart to believe that a solution

---

40 TNA, AIR 37/1249, 21 Army Group, Operation Market Garden, 17-26 Sept 1944.
41 Ritchie, *Arnhem*, 120.
42 TNA, AIR 37/1214, 1st British Airborne Corps – Allied Airborne Operations in Holland Sept-Oct 1944.
43 TNA, WO 171/393, 1 Airborne Div. Op Instr No.9, Confirmatory Notes on GOC's Verbal Orders, 1st Airborne Division War Diary, September-December 1944.
44 Cirillo, *Market Garden and the Strategy of the Northwest Europe Campaign*, 45-46.
45 Ritchie, *Arnhem*, 249-261.

to the difficult operational problem he had been set was possible. This is where his personal qualities are likely to have affected his assessment of the situation and the likelihood of success for the mission.

Urquhart would have been encouraged by the (mis)application of his previous successful experience in command of an infantry brigade; he did not fully comprehend the unique problems faced by an airborne formation. This would have been reinforced by the external social influence pressures he faced. By way of a recap, the social influence pressure of Authority was exerting pressure on him; he had been given a direct order and had no real choice but to comply with it. There was also group pressure or Conformity; around him nobody was seriously challenging the plan. Furthermore, both Urquhart and his senior officers, indeed the whole division, were eager to get into action using the plan because of the previously cancelled operations; this is the principle of Commitment and Consistency at play. Finally, as we have seen, the window of opportunity to use the division in the airborne role, before possible disbandment, was rapidly closing due to the lateness of the season (Scarcity principle). These are strong influencers that Urquhart would have found difficult to resist. The general sense of victory euphoria that existed at the time and the ambiguously optimistic intelligence being fed to him by higher authorities also added extra support to him adopting a positive appraisal of the viability of his plan. Let's return to his Headquarters.

Urquhart's Headquarters, Grantham, England, 10th September 1944

There were positive aspects to the issues Urquhart faced. Firstly, although the landing and drop zones were a long way from his division's objectives, they were, ideal for his purposes. The selected zones situated in open country to the west-north-west of Arnhem were comprised of large areas of flat ground, screened by belts of woodland. As far as he could see, the selected zones had clear advantages, as he would later state in his memoirs: 'the ground would be firm and also ideal for quick re-grouping'.[46] (There is a strong element of truth in his view as the landing and drop zones were suitable – they were just a long way from the objectives). He could, however, overcome the distance of the landing and drop zones from the objective and the lack of a *coup de main* attack by acting on Brigadier Gerald Lathbury's (Commander, 1st Parachute Brigade) suggestion of using the armed jeeps of the Reconnaissance Squadron

---

46 Urquhart, *Arnhem*, 7.

under Major Freddie Gough to race ahead and seize the bridges.[47] He was confident that this would solve the problem of surprise.

Urquhart was also encouraged by the considerable resources available to him, after all, he was in command of an elite airborne division. Not only that, with the addition of extra air landing elements, glider pilots and the Polish brigade, his command was nearly double the size of a normal airborne division. To add to this, the Market Garden plan also called for the 52nd Lowland Division, which was air-portable, to be flown into the nearby Deelen airfield once seized.[48]

There was also the assurance, that as part of the overall Market Garden plan, XXX Corps would reach his division within two days.[49] Thus, he was told, regardless of his concerns about concentrating sufficient combat power and the locations of the landing and drop zones, the operation would be over within 48 hours and thus the issue of the last brigade dropping on the third day was an irrelevance.

Urquhart's Doubts

Urquhart was unsure of what to make of this assurance. His memoirs show some ambivalence. He would later state in his autobiography that it seemed 'hardly over-optimistic' that, given the speed of recent advances, XXX Corps would manage it. On the other hand, and of course this may just be based on the benefit of hindsight, he also states that it 'needed no military genius to detect the snags'.[50] He may have been ambivalent, but he did not dismiss this assurance out of hand at the time.

To summarise, although Urquhart was faced with various constraints on his planning, several ostensibly attractive options presented themselves to solve the problems created by the restrictions. He may also have not, given his infantry background, fully appreciated the difficulties he was facing. Looking at his mindset in Decisional Conflict terms, Urquhart would, on balance, have felt that although a tall order, there was a viable solution to be found to the problems he faced, at least in the early phases of the planning. We would suggest, therefore, that he made a positive appraisal of the second question in the DCM model; there was a viable option. Time was now the factor.

---

[47] TNA, WO 171/393, 1 Airborne Div. Op Instr No.9, Confirmatory Notes on GOC's Verbal Orders, 1st Airborne Division War Diary, September-December 1944.
[48] Middlebrook, *Arnhem 1944*, 43.
[49] TNA, WO 205/313 21st Army Group Operation 'Market Garden' Plans and Instructions.
[50] Urquhart, *Arnhem*, 22.

Time

It is time to address the final question in the Decisional Conflict Model and return to your nightmare of a holiday in Florida. You have accepted the risk; the hurricane is coming your way. In a new scenario, you do have some attractive options. There might not be any private ambulances, but plenty of rental cars are available and private nursing staff that can be hired. You might be able to combine both and drive out of harm's way. There are also various hardware stores selling hurricane kits you can use to fortify the hotel room if you decide to stay put. Your answer to the second question in the model then, is positive, there are viable solutions, it just needs some thinking about.

According to the DCM model, having addressed the feasibility issue, you then move on to the third and final question; do you have time to search for, develop and implement a workable plan? If your answer to this question is positive, you do have time, you will then engage in purposeful and comprehensive information-processing. This is the rigorous approach to problem-solving referred to as Vigilant information-processing that we discussed in the last chapter. Your thinking in this (fifth) condition can be complex (in Integrative Complexity terms) as you thoroughly search for and analyse the available information, generate alternative options, compare these in terms of costs and benefits and finally reach a considered and well thought through conclusion.[51]

If, however, your assessment is that there is insufficient time available, you will adopt the fourth coping pattern of Hypervigilance. In this state, you will engage in problem-solving activity, but your information-processing, because of the perceived short timescale, is rushed, unstructured and incomplete. You rapidly search for a way out of the dilemma, shifting back and forth between alternatives, and impulsively seize upon a hastily contrived solution that seems to promise immediate relief. You take shortcuts, miss things and perhaps rely on experience to solve the problem, what worked last time? If you are in the Hypervigilance pattern, you will meet the criteria for vigilant information-processing but to a limited and variable extent.[52] The quality of your information-processing is, therefore, dependent on the amount of time you perceive as available to reach a decision and act, the shorter the timescale the worse your information-processing. The next

---

51 Janis & Mann, *'Emergency Decision Making'*, 35-48.
52 Ibid., 35-48.

question we need to consider, then, is did Urquhart have enough time to come up with a workable plan?

Urquhart and his planners were under a lot of pressure, time pressure especially. Operation Market Garden was conceived and planned within a short period, albeit the plan extemporised on a previous operation, Comet. The operation was given the go ahead on 10th September and was planned to begin on 17th September.[53] First Airborne planners, Urquhart included, therefore had six days to prepare for the drop into Arnhem, and even less to plan.

Urquhart's Timeframe

Urquhart called his first Orders Group to brief his plan for the division's role in Market Garden at 1700 hours on 12th September.[54] It had been two days since Urquhart had himself attended a briefing for Market Garden at Browning's Headquarters at Moor Park, late on 10th September. He had received his written orders a little afterwards.[55] His planning had, therefore, been essentially conducted over a 48-hour period. Urquhart states in his autobiography that '[T]he plan had to be produced quickly'.[56]

Hypervigilance

If we examine Urquhart's assessment of the time available to him to plan effectively, there are mitigating factors in conflicting directions. Urquhart was an experienced staff officer, and his divisional staff had had lots of practice; as he states in his memoirs: 'They [the 16 previously cancelled operations] had the effect, however, of … advancing our planning to a fine art …'.[57] He also had an existing template to work from in the form of the plan for Comet. Although he had a head start, Urquhart was forced to work within timescales that were compressed. Time was tight with a lot to do; his brigades would be jumping into the Arnhem area only, rather than across the whole corridor, including both Eindhoven and Nijmegen. This was a different challenge. Furthermore, his restricted access to the available intelligence meant that he was not able to consider all the different factors affecting the situation.

---

53 TNA, WO 171/366 1st Airborne Corps War Diary: Operation Market Instruction #1, 13 September 1944.
54 TNA, WO 171/393 HQ 1st Airborne Division War Diary, 1st-30th September 1944.
55 TNA, WO 205/313 21st Army Group Operation 'Market Garden' Plans and Instructions.
56 Urquhart, *Arnhem*, 9.
57 Urquhart, *Arnhem*, 18.

What does this mean for how he would have answered the final question (in the DCM model) about the availability of time to properly plan? The tight timescale meant his answer, we suggest, would have been in the negative. The impact of this assessment means he would have felt the need to cut corners and work with assumptions on certain issues during the planning process, something not unusual but does lead to the possibility of making key errors. He would, therefore, as he worked through the planning process, have adopted the Hypervigilance coping pattern and engaged in rushed and incomplete information-processing, drawing on his past experiences to find solutions. The impact of this approach, in term of the errors he committed will be explored in the next chapter, but before we turn to this, there is an interesting twist, one that takes us back to the incident described at the start of this chapter; something happened to shift Urquhart from his initial optimistic view to his fatalistic comment about a suicide mission.

Reappraisal

The exact nature of Urquhart's information-processing and decision-making will be explored in more detail in the next chapter, we will restrict ourselves here to the process outlined in the Decisional Conflict Model. In his initial Hypervigilant state, Urquhart did attempt to solve the problem with which he was presented.

Urquhart's Headquarters, Grantham, England, 15th September 1944

By 15th September Urquhart had tried to find solutions to the different problems he faced but had drawn a blank on each occasion. He had asked Browning for 40 more aircraft but was refused on the grounds the needs of the two American divisions had to be prioritised as Market Garden was a bottom to top operation.[58] He had also questioned Browning's need to use 38 gliders to bring in his Corps Headquarters, which had no direct role to play in the combat zone, arguing that he could make better use of them, this was refused also.[59] Urquhart had tried and failed to get landing and drop zones closer to the bridges.[60] He also wanted to insert his division on both sides of the bridges, but this again had been rejected by Hollinghurst and the air

---
58 Harclerode, *Arnhem*, 50.
59 Bennett, *A Magnificent Disaster*, 39.
60 Hibbert, *The Battle of Arnhem*, 41.

planners on the basis of threats to the transport aircraft ostensibly posed by the flak batteries based at the nearby airfield at Deelen.[61] In a similar fashion, he had also asked for two lifts on the first day so as to be able to insert his entire division; this had also been rejected, due to concerns about lack of transport aircraft, crew fatigue and aircraft maintenance.[62] Everything he had tried, every argument he had put forward, had failed.

Urquhart's Acceptance

Urquhart did, therefore, raise concerns as he conducted his planning for Market Garden, it is not clear why Browning did not address these issues.[63] There is some criticism over Urquhart being too accepting of the RAF warnings about flak in the area.[64] Bennett has argued that Urquhart may have got his way if he had pushed harder. Browning did consult Richard Gale of 6th Airborne Division about the plan for dropping 1st Airborne. In response Gale had stated at least one brigade needed to be dropped near to the bridges; an issue, he claims in retrospect, he would have resigned over.[65] There is further criticism over Urquhart being too concerned about having a safe approach to the landing and drop zones and too focused on achieving a tidy insertion for his division.[66] The key point is that Urquhart failed in his attempts to solve the problems he faced. What was the impact of this failure on his mindset?

Dynamic Model

The DCM model is dynamic in nature; it views individuals re-appraising the antecedent conditions as events unfold.[67] Let's return for one final time to your Florida dilemma. In this changing scenario, you were looking at getting hurricane kits from hardware stores or renting a car and hiring a private nurse to help you drive out of harm's way. As you work through these options, it becomes clear that all the kits are sold out. You might be able to

---

61 TNA, AIR 37/1214, 1st British Airborne Corps-Allied Airborne Operations in Holland Sept-Oct 1944.
62 Bennett, *A Magnificent Disaster*, 50.
63 Bennett, *A Magnificent Disaster*, 40.
64 Hibbert, *The Battle of Arnhem*, 42.
65 Bennett, *A Magnificent Disaster*, 40.
66 Hibbert, *The Battle of Arnhem*, 39.
67 Janis & Mann, 'Emergency Decision Making', 35-48.

rent a car, but there are no private nurses available. It turns out that neither of these two options are actually viable. As you reassess the situation it becomes clear that you and your partner are at risk, but there is no workable solution. Your reappraisal of the situation means you slip back into a Defensive Avoidance coping strategy; pulling the duvet over your head and hoping for the best is probably the only option. How did Urquhart react to the changing circumstances he faced?

Urquhart's Bolstering

As Urquhart was foiled in his attempts to solve the problems he faced, he is likely to have re-appraised his evaluation of the second antecedent condition (whether there was a viable solution) and come up with a negative answer. This means that, at some point he would have shifted to Defensive Avoidance. This would have taken the form of either: Procrastination (failing to come to a decision); Scapegoating (shifting the responsibility for making the decision to someone else); or Bolstering (accepting the situation, settling on the least objectionable alternative and boosting its attractiveness by wishful thinking). Urquhart does not seem to have vacillated in deciding (Procrastination), he issued his orders within 48 hours. He also did not look to another to shoulder the responsibility (Scapegoating); he drew up the plan and took responsibility for implementing it. It does appear, though, that later in the planning process, he adopted the coping pattern of Bolstering; he accepted the constraints, coming up with the best plan he could, hoping that he would in fact face little opposition, that the armed jeeps would seize the road bridge, his battalions would be able to march quickly to reinforce them and he would be relieved within two days.

Ultimately, although making the most of the best option available, Urquhart would also have accepted that the mission was very difficult if not (nigh on) impossible. This returns us to the incident reported by Browning's ADC, Captain Edwin Newbury at the start of this chapter and Urquhart's remark about a 'suicide mission'.[68] Although only an isolated incident, with one witness, if true, this suggests that having essentially completed the planning process (in a Hypervigilant manner), Urquhart did, eventually, shift towards Bolstering as a coping mechanism.

---

68 CRCP 108/5, Cornelius Ryan Collection of World War II Papers.

Conclusion

In summary, Urquhart's interpretation of the three antecedent conditions in the Decisional Conflict Model indicates that the coping pattern adopted by him during the planning process was one of Hypervigilance shifting to Bolstering at the end. In terms of the first question, the level of risk prevalent in the situation, he was aware that elements of two Panzer Divisions were in the Arnhem-Nijmegen area, but their level of combat effectiveness was downplayed to him. He would have perceived sufficient risk to experience decisional conflict and move on to consider the next question in the process. He was clearly faced with a difficult problem; the air plan imposed severe constraints on his planning. Conversely, the intelligence picture presented to him of the poor fighting quality of German troops in the area, the resources available to him and the assurances his division would be relieved within two days, coupled with the social influence pressures meant that he would have initially seen these as challenges that could be overcome. Having believed a viable solution was possible, he had tight timescales to work within. Essentially, he had 48 hours to formulate his approach (albeit one based on the previous Comet plan). His assessment of these different factors would, therefore, have led him to adopt the DCM Hypervigilant coping strategy. As his attempts to solve the problems he faced failed, a *coup de main* attack was refused, the landing and drop zones remained a distance from the bridges and his division inserted across three days, he reappraised the situation and then shifted to a Bolstering stance towards the end.

Other than his late shift to Bolstering, Urquhart spent the crucial period of the planning process in the Hypervigilant coping pattern. In this condition, whilst actively engaged in the planning for the operation, his information-processing and decision-making activities would have been rushed, unsystematic and incomplete. This curtailed process would have precluded extensive information search and problem-solving and would have led to a reliance on more heuristic-based judgements and simplistic solutions, drawing on his own experiences when developing his plan. In the next chapter we will examine the extent to which the effect of his Hypervigilance coping pattern impacted on his operational planning.

14

Urquhart's Action – Framing

'Perhaps our ideas were wrong'.[1]

Introduction

Browning's Headquarters, Nijmegen, Holland, 26th September 1944

Urquhart, although utterly exhausted by the ordeal of the last nine days, lay awake in the bed provided for him at Browning's Headquarters after his evacuation across the Rhine. He had just overseen the virtual destruction of his division, bringing less than a quarter of his men back with him. It was a bloody disaster; what went wrong?

Urquhart's Issues

As discussed in the previous chapter, there is evidence to suggest that Urquhart adopted the Hypervigilant coping strategy when planning for his division's role in Market Garden. Hypervigilance involves curtailed or only partially effective information-processing. This means that he would be more likely to engage in heuristic-based, System One-type thinking and thus be more prone to cognitive biases and errors. This chapter examines his actions to explore whether he committed these mistakes when planning for Operation Market Garden. First, we need to examine his plan.

Urquhart's Plan

Urquhart's options had been constrained by the shortage of aircraft and the air planner's stipulation of one lift each day. He had decided to drop 1st

---
1 Urquhart, *Arnhem*, 199.

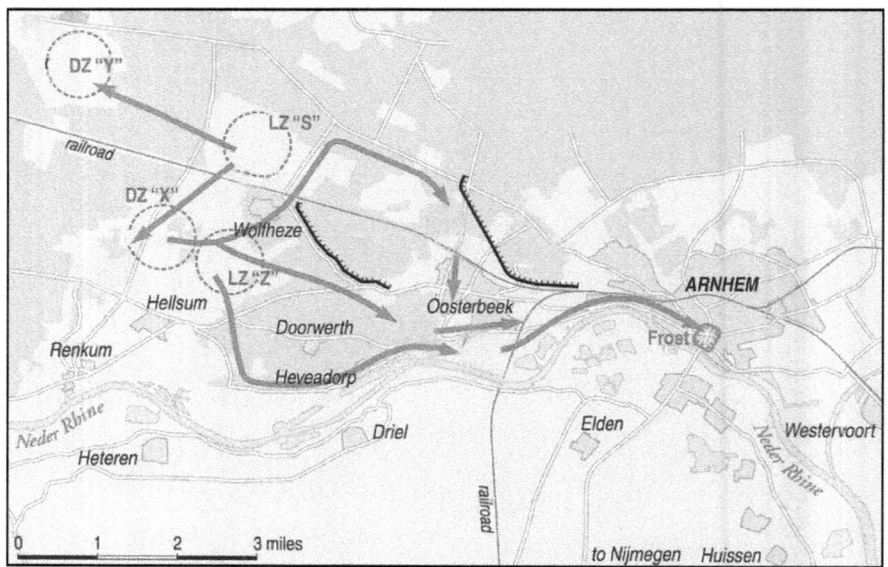

Figure 14.1 Landing and drop zones, and routes into Arnhem

Parachute Brigade and most of 1st Airlanding Brigade, plus some HQ and divisional troops on day one. First Parachute Brigade had been tasked with capturing the division's main objectives, the road, rail and pontoon bridges at Arnhem. He had tasked 1st Airlanding Brigade, the strongest formation in the division, to hold the landing and drop zones (illustrated in Figure 14.1) until the second day, then move east to form the western perimeter of the bridgehead around Arnhem. The rest of 1st Airlanding Brigade would arrive on the second day along with 4th Parachute Brigade, who would form the northern sector of the bridgehead.[2]

First Polish Independent Brigade would drop to the south of Arnhem on the third day and take up positions to the east of the town.[3] This meant that only one brigade was available to seize the objectives on the first day. In the absence of a true *coup de main* operation, and to compensate for the long advance to the bridges, Urquhart had agreed to Lathbury's suggestion of

---

2 TNA, WO 171/393, 1 Airborne Div. Op Instr No.9, Confirmatory Notes on GOC's Verbal Orders, 1st Airborne Division War Diary, September – December 1944.
3 Ibid.

using the Divisional Reconnaissance Squadron, equipped with armed jeeps, to move ahead of the main body of advance to seize the bridges.[4]

Criticism

This was the basic plan for the division, as Clark has pointed out, it was based on the 'if all went well' principle.[5] Several criticisms have been levelled at Urquhart's plan; in private, Hackett believed, at the time, that the plan was unprofessional.[6] Dempsey has accused Urquhart and his staff of 'inept planning'.[7] Some of the participants, notably Urquhart and John Frost (commanding 2nd Battalion of the 1st Parachute Brigade), have themselves, with hindsight, identified various problems with the plan.

We would argue that these issues stemmed partly from Urquhart's adoption of the Hypervigilant coping strategy which in turn drove lower levels of Integrative Complexity. A key question is how did this low Integrative Complexity manifest itself in Urquhart's planning? Before we get into these specific problems, it will be useful to look at the bigger picture and Urquhart's general approach to the operation; in this he was prone to similar biases that we have seen with Montgomery and Browning. In a similar manner to Browning, he certainly had a lot invested in the operation.

Sunk Cost Effect

The Sunk Cost Effect occurs when you have difficulty stopping an activity or giving up on something when you have a lot invested in it, whether it be time, energy, money, or indeed all three. The investment itself becomes your reason to carry on; the greater the investment, the greater your urge to continue.[8] Urquhart had been in command of the division since the start of the year. He had put a lot of effort into preparing it for operations and was frustrated by the lack of action. This effort and frustration had created a strong momentum behind the operation, as evidenced by Urquhart's comment about his division being 'battle-hungry'.[9] Urquhart had a lot invested in his division and a lot to

---

4 TNA, WO 171/393, 1 Airborne Div. Op Instr No.9, Confirmatory Notes on GOC's Verbal Orders, 1st Airborne Division War Diary, September – December 1944.
5 Clark, *Arnhem*, 106.
6 Bennett, *A Magnificent Disaster*, 241.
7 TNA, WO 285/29, Dempsey to Ellis, 18 June 1962.
8 Plous, *The Psychology of Judgment and Decision Making*, 243-244.
9 Roy Urquhart, *Arnhem*, 18.

lose if this operation didn't go ahead; this also impacted on how he viewed or framed the situation.

Framing Effect

We have already examined the impact of the Framing Effect on Montgomery and Browning. Framing the situation in terms of potential losses typically leads people to more risk taking. We have also discussed at length what Urquhart stood to lose if Market Garden did not go ahead, this included the potential 'loss' of his division not being deployed in the airborne role before the war ended, and even being disbanded or being deployed in the line infantry role.[10] The Framing Effect was, therefore, most likely to be impacting on Urquhart and affecting his risk assessment and in particular, how he viewed the fighting capacity of the German forces in the area. His risk assessment would also have been affected by another bias.

Exposure Effect

The Exposure Effect is a psychological mechanism where the regular exposure to a situation or person desensitises you to that situation or person. This desensitisation has the effect of making the situation or person feel less risky or threatening.[11] The 16 previous operations, some of which had been cancelled at the very last minute, had created a greater acceptance of the risks involved with the operation, with Urquhart himself commenting on the 'tendency to take light-heartedly the less encouraging factors'.[12]

Overall, in terms of his risk perception, Urquhart was encouraged to be more inclined to accept the risks inherent in the operation. Worn down by the repeated cancellation of operations (Exposure Effect) Urquhart was pushed into accepting the risk. This tendency was exacerbated by him focusing on what he might lose (Framing Effect), by not getting his division into action; this was something he was heavily invested in and didn't what to miss out on, he had sunk too much into it (Sunk Costs). The downplaying of the risk also led to a dangerous overconfidence.

---

10 Michael Carver, *The Seven Ages of the British Army*, 277.
11 Development, Concepts & Doctrine Centre, *Understanding & Decision Making* (Shrivenham: Joint Doctrine Publication 04, 2016), 68.
12 Urquhart, *Arnhem*, 18.

Optimism Bias

Urquhart had a great deal of confidence in his division's fighting ability, he was not alone in this; his confidence was shared more widely. By September 1944, the Allies placed a lot of faith in airborne forces, but it was perhaps misplaced. Sebastian Ritchie argues that this was partly based on a misperception of the success of the D-Day airborne operations.[13] There was an overconfidence within the ranks of the Airborne Forces in general. It is worth revisiting a couple of previous observations. Urquhart states: 'Many of the chaps had done well in North Africa and in Sicily. But, in places, there was a slight reluctance to accept that further training was vitally necessary'.[14] This view is further corroborated by Major Tower's comment that the 'airborne boys … overestimated their prowess'.[15]

Martin Middlebrook has commented on the fact that by the time 1st Airborne went into action in September 1944, it was keen but stale.[16] Otway points out that 1st Airborne had seen eight days of fighting in Sicily in July 1943, but had not seen any action since then.[17] Buckingham is of a similar view, stating that it was, to some extent, not fully battle worthy. He cites as evidence the fact that 1st Parachute Brigade did not perform well in the crucial initial hours on the first day of the operation.[18] There was, as Ritchie points out, a certain lack of urgency. For example, when 3rd Parachute Battalion stopped for the night when it was blocked trying to get into Arnhem, Lathbury, the brigade commander, was not unduly worried.[19] This brigade was not the only issue, the problems were division wide as Middlebrook points out, not only did it have a commander with no airborne experience, it had not actually fought or indeed to all intents and purposes trained, as a division before.[20]

Buckingham describes 1st Airborne as a loose collection of formations rather than a cohesive unit.[21] There were problems at battalion and brigade level and the Divisional Headquarters had only been in operation once; the division had only operated as a single entity twice, once operationally and

---

3 Ritchie, *Learning to Lose?*, 25.
4 Baynes, *Urquhart of Arnhem*, 72.
5 Middlebrook, *Arnhem 1944*, 22.
6 Ibid., 39.
7 Otway, *Airborne Forces*, 263-264.
8 Buckingham, *Arnhem 1944*, 50-52.
9 TNA, WO 171/393, 1st Airborne Division War Diary, September 1944, Annexure N, Operation Market, Story of 1 Parachute Brigade.
20 Middlebrook, *Arnhem 1944*, 39.
21 Buckingham, *Arnhem 1944*, 51-52.

once in training, both periods for only a week.²² This lack of cohesion would bear a grim fruit at Arnhem; the battalions of 1st Parachute Brigade would act independently of each other on discrete tasks. A multi-battalion attack that was put in on 19th September, to try to force a way to the bridge, was poorly directed and ultimately unsuccessful.²³ First Airborne, therefore, suffered from a lack of experience and cohesiveness. This meant that, by September 1944 the division, as Clark suggests, was 'not only stale, it also ended up distinctly undercooked'.²⁴ This would lead to problems for the division in relation to Market Garden, but it's time to return to the plan. The first issue we need to consider is, perhaps, the most well-known and certainly the most contentious, the selection and location of the landing and drop zones.

Drop Zones Too Far from Objectives

Browning's Headquarters, Nijmegen, Holland, 26th September 1944

Urquhart, in his After-Action-Report on the operation, identified that the landing and drop zones were too far away from the division's objectives, therefore surprise was effectively lost.²⁵ Market Garden probably failed in the first 24 hours because 1st Airborne failed to get the bulk of the division to Arnhem bridge due to the distances from the landing and drop zones.²⁶ The planning rule, enshrined in airborne doctrine at the time, was that the zones should be no more than five miles from the objective; whereas at Arnhem, some were nine miles distant.²⁷ Gavin, when Urquhart had outlined his plan at a briefing session, had been horrified turning to a colleague to say, 'my god he can't mean it'.²⁸ Kurt Student, who commanded German airborne forces throughout the war, agrees, stating afterwards the main problem was British landing and drop zones were too distant from the objectives.²⁹

---

22 Ibid., 24.
23 Ritchie, *Arnhem*, 185.
24 Clark, *Arnhem*, 116.
25 Urquhart, *Arnhem*, 198.
26 Buckingham, *Arnhem 1944*, 90.
27 AHB, 1st Airborne Division Report on Operation Market, 10 January 1945.
28 Buckingham, *Arnhem 1944*, 83.
29 Harvey, *Arnhem*, 43.

Culpability

The selection of the zones is a hotly contested issue. One of the dominant narratives about Arnhem is that the locations of the zones were imposed on Urquhart by the RAF.[30] There had, though, not been any real argument over their location, Urquhart had only preferred other zones. He had not raised major objections at the time, the official records show that there had not been any disagreements between senior officers.[31] Indeed, as we have seen, Urquhart chose the drop and landing zones at Arnhem, but within the restrictions placed upon him by Hollinghurst, the RAF planner.[32] The outcome of all of this was a loss of surprise, this would not have been a problem if the division had faced little resistance. He had, however, been a little too optimistic on that front.

Confirmation Bias

We have already explored Confirmation Bias in our discussion on Browning, so we will not labour the point here; Urquhart appears to have been affected in the same way. Confirmation Bias is the tendency to interpret information in line with a preconceived idea. This is where we come back to the intelligence picture. The prevailing wisdom within the Allied camp at the time was that the Wehrmacht was crumbling and incapable of organised resistance. This victory euphoria sets the backdrop to the preconceived notion that German resistance at Arnhem would be weak, hence Urquhart's comment to Sosabowski 'But there will be no heavy German resistance'.[33] The underplaying of the fighting strength of II SS Panzer Corps, stating that the troops would not be capable of any 'organised resistance',[34] served to confirm this notion, thus reinforcing the effect of the Confirmation Bias. Another outcome of the positioning of the landing and drop zones and the need to keep open the routes into town, was that 1st Airborne struggled to muster sufficient combat power to fight through to the bridges in sufficient numbers.[35]

---

30 Buckingham, *Arnhem 1944*, 90.
31 Ritchie, *Arnhem*, 178-179.
32 Baynes, *Urquhart of Arnhem*, 94.
33 Sosabowski, *Freely I Served*, 146-147.
34 WO 171/393, 1 Airborne Division Report on Operation Market, Part V, HQ 1st Airborne Division War Diary, 1st-30th September 1944.
35 Harvey, *Arnhem*, 9.

Lack of Combat Power

With the air lift spread across three days and the need to hold the landing and drop zones for these later lifts, at the outset, only one third of the division was available to seize the bridges.[36] The irony of this was the use of one brigade to take the objectives at Arnhem was essentially the same plan as Comet, which was cancelled due to the stiffening German resistance that the upscaling to a whole division was meant to address.[37]

This meant that out of 40 rifle companies in the division, only 20 actually dropped on the first day with 11 of these required to stay back to hold the landing and drop zones, therefore less than a quarter of Urquhart's infantry strength was deployed to advance on the bridges.[38] Urquhart also probably left too much of 1st Airlanding Brigade behind to protect the zones, thus contributing to the weakness of the assault force.[39]

John Frost would later raise the question of the requirement to drop 4th Parachute Brigade on the second day in the same area that was used on day one. This meant that, in his words: 'the strongest brigade of the division [1st Airlanding Brigade] was tied down securing the D.Z. Y for the 4th Brigade'. Frost goes on to suggest: 'if it was going to be acceptable for the Poles to drop south of the river near the bridge on D plus 2, it was surely worth taking the risk of putting the 4th Brigade down there on D plus 1'. This option would have had the benefit of releasing all of 1st Airlanding Brigade advance to the bridges which 'if this had been properly handled, even despite the presence of the II SS Panzer Corps, they would have been ensconced in and around the town during that first night, and on both sides of the river'.[40]

Greater combat power could have been generated by dropping more infantry and less divisional troops (especially artillery) in the first air lift. This is what Gavin decided to do with the 82nd drop at Nijmegen. Urquhart stuck with established British Army doctrine regards his force composition on the first day by including more divisional troops. In a similar vein, 3rd Battalion could have been committed upfront rather than held back in reserve, but this would have run counter to established British military practice. Both these options would have produced greater combat power but would have required greater flexibility in thinking to deviate from established doctrine.

---

36 Middlebrook, *Arnhem 1944*, 17.
37 Buckingham, *Arnhem 1944*, 94.
38 Harvey, *Arnhem*, 43.
39 Baynes, *Urquhart of Arnhem*, 164.
40 Frost, *A Drop Too Many*, 258-259.

The brigade's efforts were also not helped by a plan that dispersed its efforts across too many objectives.[41] A harsh critic would probably suggest the plan was 'typically airborne' in nature, in that it was complicated and wasted its combat power on peripheral tasks.[42] The brigade's effort was probably unnecessarily distracted by the effort to seize not just the road bridge but also the pontoon and rail bridges at Arnhem.[43] In contrast, Taylor, faced with a similar problem, had managed to get the number of his 101st's objectives reduced.[44] First Parachute Brigade's effort at Arnhem had also been further dispersed by the need to keep the routes into town open.[45]

Overall, Urquhart's lack of airborne experience meant he perhaps did not factor in strongly enough the need for the battalions to move quickly in strength towards the bridges; he may have relied too much on his infantry experience.

Anchoring Effect

Having previous experience of dealing with a problem can help you in your problem-solving as it can obviously provide a ready-made solution or at least an analogous template to work from. Problems can arise, of course, if that experience 'anchors' thinking and you fail to make sufficient adjustments to the new situation. We discussed in Chapter Twelve that Urquhart was, in Jaques' Stratified Systems Theory terms, a Mode Four. This suggests that Urquhart, faced with a requirement to solve a novel and complex problem (Level Five within Jaques' model), was predisposed to apply his previous knowledge and experience to the situation, the problem was his previous experience was not airborne in nature.

As we have seen, only half the troops inserted over the first two days were infantry. This is in stark contrast to the two American divisions who, with the emphasis on speed and the need to seize multiple objectives, brought in a very high proportion of infantry.[46] Maxwell Taylor, for example, heavily favoured his paratroops rather than the glider-borne support echelons.[47] Urquhart, on the other hand, decided to bring in a greater proportion of

---

41 Ritchie, *Arnhem*, 184.
42 Buckingham, *Arnhem 1944*, 95.
43 Frost, *A Drop Too Many*, 201.
44 Powell, *The Devil's Birthday*, 72.
45 Bennett, *A Magnificent Disaster*, 228.
46 Harclerode, *Arnhem*, 56.
47 Clark, *Arnhem*, 103.

divisional support troops such as artillery.[48] There is, therefore, a stark contrast between Gavin's and Taylor's approach on the one hand and Urquhart's on the other. It is difficult to argue against the contention that the two experienced airborne officers, understanding the need to move quickly, emphasised infantry, whereas the experienced infantry officer opted for the more traditional balance of support arms.

Overall, the lack of combat power was a significant problem. It led to the division having difficulty fighting its way through tough German resistance on the way to the bridges. This meant Arnhem bridge was not held in any strength. There was also another issue with seizing the bridge, the lack of a *coup de main* operation and Urquhart's solution to this problem.

Misuse of Reconnaissance Squadron

It seems clear now that there should have been a *coup de main* attack on the bridge;[49] indeed, the original Comet plan included such an assault. Urquhart had raised an objection about this, both he and Chatterton had unsuccessfully petitioned the RAF to state they would be prepared to take the increased casualties of a glider-borne *coup de main* operation to achieve surprise. Chatterton had also gone to Browning to ask for a *coup de main* operation but had been told it was too late.[50] This had not been a popular move within the division, Chatterton had been accused of being a 'murderer' and an 'assassin' for championing just such an operation.[51]

Urquhart had, in the absence of a true *coup de main* operation, agreed to Brigadier Lathbury's proposal to use the Divisional Reconnaissance Squadron, equipped with armed jeeps, to race ahead of the main body of advance to seize the bridges.[52] This was, perhaps, a misuse of the division's reconnaissance capability as it had essentially left Urquhart blind.[53] This problem had manifested itself when the division became stalled due to the stiff German resistance. Freddie Gough, commander of the squadron was himself unsure of this use of his reconnaissance jeeps; when given this new tasking, he had asked for light Tetrach airborne tanks, but these were not

---

48 Powell, *The Devil's Birthday*, 61.
49 Frost, *A Drop Too Many*, 257.
50 Buckingham, *Arnhem 1944*, 85-86.
51 Baynes, *Urquhart of Arnhem*, 94.
52 TNA, WO 171/393, 1 Airborne Div. Op Instr No.9, Confirmatory Notes on GOC's Verbal Orders, 1st Airborne Division War Diary, September – December 1944.
53 Frost, *A Drop Too Many*, 256.

forthcoming.⁵⁴ The jeeps were, also, not really suited to the task and would have been more useful in acting as a communications relay when the radio links became problematical.⁵⁵ As events turned out, the jeeps were ambushed en route to the road bridge and so had failed to reach their objective.⁵⁶ Further, this meant the jeeps had failed to fulfil their task as either a *coup de main* operation or a reconnaissance function. They had though, to Urquhart's mind, been an attractive option.

Availability Heuristic

The Availability Heuristic, as we saw in the discussion of Montgomery's biases, is your tendency to recall something that is more vivid, easily recognisable or stronger in your memory. The problem is the thing you picture, might not be suitable or apposite. We would argue the imagery associated with the armed jeeps, racing ahead of the division to seize the bridges was just too attractive and vivid to resist. They could easily be pictured hurtling down the main roads into the town shooting their way through any opposition. We would also argue that a more flexible and effective plan would have been to hold back 1st Parachute Brigade whilst the Reconnaissance Squadron reconnoitred the main roads to determine where opposition was lightest and then to have fed the battalions along that route. This would of course, require a greater degree of flexibility in planning, which brings us on to the next problem, the fixation on the main routes into Arnhem.

Fixation on Main Roads

There were three main roads that led from the area of the landing and drop zones to the bridges; these had been an attractive option for getting into Arnhem quickly. Urquhart and Lathbury had opted to push each of the three battalions of 1st Parachute Brigade down these roads, one per road.⁵⁷ Brigadier Lathbury had given 2nd Parachute Battalion the task of proceeding along the lower road into Arnhem to seize the bridges, occupying positions on both banks of the river. Third Battalion had been tasked with moving along the main or higher road and then holding the western part of Arnhem.

---

54 Baynes, *Urquhart of Arnhem*, 98.
55 Harclerode, *Arnhem*, 165.
56 Powell, *The Devil's Birthday*, 64.
57 Powell, *The Devil's Birthday*, 64.

First Battalion had been tasked with occupying the high ground to the north of Arnhem.[58]

The problem, as it turned out, was the selected routes into Arnhem were tree lined roads running through an urban area, this had proven to be excellent terrain for defence and ideal cover for ambushes.[59] German opposition on these routes held up the advance and imposed crucial delays. This would have been less of a problem if the brigade was able to skirt round and outflank these obstacles; this was not the case. Urquhart and his planning staff had been too focused on the main routes into Arnhem and had failed to consider other options, such as the side roads. He would later state in his memoirs: 'We should have taken more short cuts. Our main preoccupation was to push along the main roads, and there were not many of these'. This meant that the advance was halted as soon as opposition was encountered because: 'The high garden walls and wire fences made movement off the roads a laborious business'. Urquhart himself noted that 'we failed to make enough use of the side roads also because nobody knew where they led, and we were short of town maps'.[60]

The difficulties imposed by using the main roads had been exacerbated by the failure in the wireless communications, as Urquhart would later comment: 'Yet even the restricted movement produced by the Arnhem Road system and our failure to use the secondary roads adequately would not have prevented a switch of some units – if our wireless communications had worked better'. This facility would have meant '3rd Battalion could have been switched to the lower road earlier'.[61]

These issues could have been, in part, mitigated, through consultation with his Dutch liaison officers, but Urquhart had not taken advantage of this. He would later acknowledge: 'We could have made more headway if we had ... taken more advantage of the Dutch liaison officers attached to the division'.[62] All in all, the rushed nature of his planning meant his information search was not as extensive as it could have been, and, indeed, was probably ended too quickly.

---

[58] TNA, WO 171/393, 1 Para Bde OO No. 1, 1st Airborne Division War Diary, September – December 1944.
[59] Powell, *The Devil's Birthday*, 64.
[60] Urquhart, *Arnhem*, 199-200.
[61] Urquhart, *Arnhem*, 200.
[62] Ibid., 200.

Premature Closure

One of the outcomes from the adoption of the Hypervigilance coping strategy is the premature closure of the search for information. The lack of consultation with the Dutch can be seen in this light. This issue was not the only problem that arose because of the lack of consideration for the ground over which the division would be operating; three other issues stand out.

Poor Terrain Appreciation

Fighting in a Built-up Area

Another factor that slowed down progress towards the objective was the urban nature of the fighting. Urquhart had to admit that the implications of this were not sufficiently factored into the planning considerations at the time. Again, he would later state that: 'There is no doubt that we came off badly in the street fighting. Most of us had not taken this problem specifically into account when the plan was made or even during the move into the town.'.[63] This was not the only problem, there are two other key issues that had a clear impact of the success of the division's mission. The first of these was the failure to identify and capture the Heveadorp-Driel Ferry located west of Arnhem.

Heveadorp-Driel Ferry

Capture of the ferry would have allowed 1st Airborne Division quick access to the south bank of the Rhine and possibly control of the rail bridge (which was blown up by the Germans). It would also have allowed Reconnaissance Squadron jeeps to have approached the road bridge from both ends on the first day as well as providing a means to by-pass German opposition en route to the bridges. Subsequently, control of the ferry (which it has been established could accommodate three Sherman tanks in one crossing), would have enabled British armour, when it arrived on 21st September to cross to the north bank. A thorough study of the available aerial reconnaissance photographs and a map of the area would, perhaps, have revealed the existence and thus potential utility of the ferry, but it had not been identified

---

[63] Urquhart, *Arnhem*, 200.

as an important feature, (a factor that the Dutch resistance, with local knowledge found inexplicable).[64] John Frost would later argue that: 'We had not been told that there was a ferry at Driel, … The failure of the planners at all levels to identify and make use of this useful asset is hard to understand'.[65] The division had paid dearly for this omission as it had for another oversight, the failure to recognise the significance of the Westerbouwing Heights, and the need to prioritise the capture of this feature.

Westerbouwing Heights

The Westerbouwing Heights, located downriver from Arnhem, was the only high ground in the area, rising 330 feet above the ferry crossing, and dominated that stretch of the river. German control of the heights during the battle's mid-stage had essentially prevented British movement in the area and the Poles effecting a crossing at Driel. The failure to recognise the importance of and to seize the heights had also created difficulties as the operation wore on. It certainly hampered movement into town for the division; it had also later prevented the Dorsets from getting across the river when XXX Corps finally arrived at the Rhine.[66] A wider ranging investigation might have highlighted the importance of the heights, but it had not, probably because Urquhart and his planning team had been too goal focussed. His division's aims were clearly set out in the operational instruction issued to him; these were to seize the bridges over the Rhine at Arnhem.[67] Urquhart would later admit in his memoirs he had been too focused on taking the bridges and had not given enough consideration to the wider, external factors.

Planning Fallacy

You will remember the Planning Fallacy bias involves a focus on internal considerations when engaged in planning to the exclusion of external factors. One indication of this is Urquhart's neglect of his Dutch liaison officers and disinterest in the information provided by the local resistance.[68] Urquhart, as we have seen, missed several external factors during his planning process, the urban nature of the environment which hampered movement as well as

---

64 Ryan, *A Bridge Too Far*, 334-335.
65 Frost, *A Drop Too Many*, 256.
66 Mead, *General 'Boy'*, 145.
67 Clark, *Arnhem*, 104.
68 Urquhart, *Arnhem*, 200.

the benefits that would have been gained by seizing the Heveadorp-Driel ferry and the Westerbouwing Heights. More consultation with the Dutch and/or a wider appreciation of the external factors might also have identified some other issues.

The other external factor that was not sufficiently addressed was the quick and fierce reaction of the German defenders; this is, perhaps, the clearest manifestation of the Planning Fallacy. This lack of consideration is evidenced by one of the most famous quotes wrongly attributed to Market Garden, Sosabowski's 'But the Germans, General … the Germans!' comment.[69] Hackett, who was present at the time, agreed with Sosabowski and has commented (about the staff at FAAA and not Urquhart per se): 'We [Hackett and Sosabowski] had both fought Germans and knew all about that. Apparently, the airborne planners did not'. Focusing on the culpability of the air planners he adds: 'Their plans to put down an airborne division were impeccable. What would happen after that was beyond them'. Borrowing a culinary analogy he concludes his damning commentary by stating: 'They were to me like cooks who prepare a superb dish and then add salt and pepper to taste. They prepared a superb deployment and then added a few Germans'.[70] Although this is not aimed directly at Urquhart, he appears to have not particularly challenged this view himself. This mindset is indicative of someone with the sort of insular thinking that is caused by the Planning Fallacy. Urquhart, as we have seen, was not working in isolation, he had his planning team and his senior officers to help and support him. We will explore the extent to which this support hindered him in the next and final section of this chapter, we have one more bias to discuss.

Groupthink

So far in this book, we have examined the decision-making for Market Garden primarily at the individual level – Montgomery, Browning and Urquhart. In the final part of this chapter, to redress the balance a little, we will examine the extent to which Urquhart's thinking and actions were affected by the group situation within which he found himself. We will explore the extent to which the group dynamic within 1st Airborne's cadre of senior officers further enabled and facilitated Urquhart's cognitive biases. We next need to look at a phenomenon called Groupthink.

---

69 Ibid., 79.
70 Harclerode, *Arnhem*, 54.

The concept of Groupthink was developed by Irving Janis who examined the decision-making of US Government officials during the Bay of Pigs disaster (the failed 1961 invasion of Castro's Cuba) and the Japanese attack on Pearl Harbour in 1941. He concluded that on both occasions, officials suffered from a collective bias he termed Groupthink. Janis states that Groupthink occurs when an overarching need for cohesiveness within a team or set of individuals results in poor decision-making.[71] Groupthink, according to Janis, is caused by several factors.

Causal Factors

One of the contributing factors that can help foster Groupthink is difficult or stressful conditions. In these instances, the group may seek to reduce the stress by reaching a decision quickly and without prolonged debate.[72] This factor was present during Urquhart's planning for Market Garden. As was discussed above, he and his subordinate commanders felt German forces in the area posed a clear risk to their division and the success of the operation; Urquhart certainly had 'no illusions about the Germans folding up at the first blow'.[73] Urquhart and his planning staff were, therefore, having to deal with a difficult situation whilst working under a tight timescale (48 hours) to develop a plan.

Another factor contributing to the development of Groupthink is the isolation of the group from external influences due to a sense of uniqueness.[74] The Airborne Forces were the elite of the British Army and saw themselves as such; 1st Airborne was no exception. Urquhart recalls upon taking up his new post: 'my impression was that airborne forces had been run rather as a family business and as a private army, and had up to date produced their own senior officers', and that 'I was soon to learn that an airborne division is a rather self-contained community into which one has to be accepted'.[75] This contributing factor was, therefore, very much in effect and so would have helped to produce the symptoms that are associated with Groupthink.

---

71 Irving Janis, *Groupthink* (Boston: Wadsworth, 1982), 174-197.
72 Janis, *Groupthink*, 174-197.
73 Urquhart, *Arnhem*, 9.
74 Janis, *Groupthink*, 174-197.
75 Urquhart, *Arnhem*, 14.

Symptoms

One of the symptoms of Groupthink is a sense of invulnerability that grows within the group.[76] This sense of invulnerability leads to an inflated sense of competence. As we have seen, there is clear evidence this was the case within 1st Airborne, with Urquhart commenting on the 'slight reluctance to accept that further training was vitally necessary',[77] and Major Tower referring to the 'overestimation of prowess'.[78]

Another outcome of Groupthink is a lack of dissent or challenge amongst group members.[79] This can be seen to be the case amongst the division and the broader airborne community by the time of Market Garden. As discussed above, having heard Richard Gale's concerns about the operation, Browning had sworn him to secrecy, silencing one experienced and informed dissenting voice.[80] Both Sosabowski and Hackett had voiced concerns about Comet, so much so that Urquhart took Sosabowski to see Browning who played down the concerns expressed. One week later, there were no dissenting voices as Sosabowski recalls the nonchalant mood displayed by his colleagues at Urquhart's briefing on Market Garden on 12th September: 'Questions were buzzing round my head, but I quickly sensed that if I started asking questions it would delay the end of the meeting; I would be unpopular with all of them and I did not think that it would be any use anyway.'[81]

Another outcome from Groupthink is a lack of critical analysis and evaluation during the group decision-making process.[82] The effect of the 16 cancelled operations and the eagerness of the Division to get into action meant that the senior commanders were not as detailed in their critical analysis as they could have been. As Urquhart himself states his team 'couldn't have cared less' and that they weren't 'thinking as hard about the risks as we possibly had done earlier'.[83] In this way, Groupthink can be seen to have contributed to the lack of Vigilant information processing discussed above.

Finally, another symptom of Groupthink is the suppression of personal doubts to maintain a sense of unanimity. We have seen the sense

---

76 Janis, *Groupthink*, 174-197.
77 Baynes, *Urquhart of Arnhem*, 72.
78 Middlebrook, *Arnhem 1944*, 22.
79 Janis, *Groupthink*, 174-197.
80 Clark, *Arnhem*, 86.
81 Sosabowski, *Freely I Served*, 145.
82 Janis, *Groupthink*, 174-197.
83 Baynes, *Urquhart of Arnhem*, 100.

of togetherness and group cohesion, so what about the personal doubts? It seems there was some surprise at Urquhart's first Orders Group.[84] In terms of individuals, Hackett for one had personal doubts. In later years, when asked to state when he first realised the operation would be a disaster, he replied, 'before it started'.[85] He claims he assumed any German response would be quick and violent;[86] and his brigade would face some hard fighting just to get to the bridge,[87] telling his people at the time this would be the case.[88] Crucially, he also felt the division 'had to get into battle almost at any price',[89] and so was not as vocal in raising his concerns as he had been.

In summary, the evidence as discussed above suggests Urquhart and his senior commanders in 1st Airborne suffered from the symptoms of Groupthink. A highly cohesive group was faced with a challenging task and short timescales. This led to minimal challenge to or critique of the planning process which in turn contributed to the underestimation of the fighting capability of the enemy and other problems. This flaw would have serious repercussions for the planning process as it meant that the constraints placed on the plan (such as the landing and drop zones being placed a long way from the objectives) were accepted and not challenged.

Conclusion

This chapter has discussed how Urquhart was prone to several cognitive biases and therefore committed several errors in his planning for Market Garden. A susceptibility to the Sunk Costs Effect, due to his investment in his division, meant he was less critical than he could have been when evaluating the risks he faced. The acceptance of the extant risk was compounded by an Optimism Bias and the Exposure Effect. The Planning Fallacy caused him to lack consideration of the external factors that threatened his plan. The potential threat of the division not being used in the airborne role or disbanded, also pushed him further to accept more risk. His reliance on his infantry experience meant he became anchored on a more traditional force composition and didn't make sufficient adjustments to cater for a novel and complex problem. The short time available to him for planning meant

---

84 Hibbert, *The Battle of Arnhem*, 39.
85 Harvey, *Arnhem*, 51.
86 Bennett, *A Magnificent Disaster*, 228.
87 Harclerode, *Arnhem*, 10.
88 Ibid., 63.
89 Middlebrook, *Arnhem 1944*, 62.

he closed his information-processing prematurely and failed to think in a wider and more flexible manner. Finally, all of this was exacerbated by the Groupthink that existed within the division. The final chapter in this part of the book explores how a Structured Analytical Technique, Outside-In View, might have helped him to think more flexibly and consider external factors.

15

Risk

Introduction

One of the issues discussed at length in the previous chapter was the tendency to be overconfident in the judgements we make. There was certainly a lot of confidence within the ranks of the Airborne Forces by September 1944. This is evidenced by Urquhart's remark about some in his division who were reluctant to accept that further training was needed and Tower's comment about the 'airborne lot' overestimating their prowess. As we discussed, one cognitive bias associated with this overconfidence is the Planning Fallacy; the tendency to focus only on internal considerations to the exclusion of external influences, especially chance factors.[1] The concerns expressed by Sosabowski and Hackett suggest that this bias was in operation and that Urquhart's planning was too internally focussed. This brief chapter uses a Structured Analytical Technique, taking an Outside-In View to explore whether this would have helped Urquhart to identify and consider the external factors that might have affected his plan.

Outside-In View

The Outside-In View technique, as the name suggests, involves viewing a problem from an external perspective.[2] As with other techniques, it follows a set of procedural steps, the first of which is to list the type of external issues that may impact on the plan. To help this process, it is useful to consider some broad themes or areas, typically the technique uses the PESTLE headings: Political; Economic; Social; Technological; Legal; and Environmental. For the purposes of this exercise, we will use headings more related to military

---

[1] Hardman, *Judgement and Decision Making*, 108.
[2] Heuer & Pherson, *Structured Analytical Techniques for Intelligence Analysis*, 228-232.

appreciations: Enemy, Weather, Terrain, Communications, Friendly Forces. The next step is to estimate what problem each issue would cause in terms of the plan; then assess the risk that this problem would pose. The final step involves deciding what can be done in terms of risk management, in simple terms this might be to terminate the risk by resolving the problem completely, treat the risk by mitigating it to some extent, or to tolerate it as a necessary evil.

As you can see, this technique is quite simple but can be quite powerful. It is often useful to draw up a simple table that sets out this analysis on one page. Let's apply the technique to what we know about the situation Urquhart was facing.

Urquhart faced several problems when planning his division's role in Market Garden. Some of these issues were known and recognised as potential problems at the time; some, he (and other commentators) identified later as mistakes that were made during the planning process. These issues were discussed at length in the previous chapter so we will only list them briefly here. The mistakes that were made or the reasons for failure at Arnhem can be summarised as: the impressive reaction of the Germans (due to their anti-airborne training and indoctrinated philosophy to counter-attack);[3] poor weather disrupting subsequent drops;[4] the over-reliance on using main roads into Arnhem;[5] the urban nature of the terrain restricting movement;[6] the failure to take and hold the Westerbouwing Heights;[7] the failure to seize the Heveadorp-Driel Ferry;[8] the breakdown in radio communications;[9] and the failure to liaise with Dutch resistance groups.[10] These are the issues that we shall use for our Outside-In View analysis. Hindsight is obviously a key issue that we again need to be mindful of here. The point of the exercise, however, is to highlight, in advance, potential issues that might arise. The use of the general planning factors, such as terrain, to guide his thinking, followed by working through the analytical process, prompting the asking of questions, might have helped Urquhart to identify these issues, and then manage the risks at the time. Table 15.1 sets out an example of the first five steps of the Outside-In View process.

---

3 Urquhart, *Arnhem*, 202.
4 Ritchie, *Arnhem*, 199.
5 Powell, *The Devil's Birthday*, 64.
6 Frost, *A Drop Too Many*, 258.
7 Mead, *General 'Boy'*, 145.
8 Frost, *A Drop Too Many*, 256.
9 Bennett, *A Magnificent Disaster*, 233.
10 Urquhart, *Arnhem*, 200.

Factor	Issue	Problem	Risk
Enemy	Reaction	Weak German units but motivated and well trained; now defending border of Germany. Trained in anti-airborne action.	Quick response; determined resistance. May well block the advance, preventing own forces reaching the bridge.
Weather	Bad weather	Lift over three days means greater probability of bad weather; disruption, delay to reinforcement and resupply.	Reduction in combat power; may struggle to hold ground for an extended period.
		Bad weather may disrupt ability to fly Close Air Support (CAS) missions.	Reduces combat power (out of range of friendly artillery). Limits Allied ability to deal with strong pockets of German resistance.
Terrain	Type	Urban nature, Arnhem is a built-up area. Good cover for enemy. Restricts own ability to manoeuvre.	Facilitates enemy defence, good opportunities for ambush. Hard to fight through. Limits ability to bypass pockets of resistance.
	Roads	Only three main roads into Arnhem from landing and drop zones.	Channels movement. Limits ability to perform flanking manoeuvres and/or bypass pockets of resistance.
	Transport	Not seizing Heveadorp-Driel Ferry loses alternative route across the Rhine.	Movement restricted to north of river, unable to assault bridge from both ends. Limits ability to outflank or bypass resistance.
	Ground	Westerbouwing Heights only high ground in area; dominates routes into Arnhem, good observation.	Enemy can observe movement and call down fire on own forces. Limits ability to move unhindered.
Communications	Radios	Historical and recent reports of problems with radios; area is wooded and built-up.	Radios fail to work. Problems of coordination within division; loss of tactical flexibility. Limits ability to direct resupply or CAS.
Friendly forces	Dutch resistance	Not consulting Dutch resistance groups loses expert local knowledge of area and German positions.	Lack of ground knowledge. Limits intelligence appreciation and restricts development of courses of action. Constrains freedom of action on ground.

Table 15.1 Urquhart's Plan: Outside-In Analysis

Risk Management

The final step of the process is to try to determine what can be done to manage the risks identified in the analysis. We have, for the purposes of this exercise, limited the managing options to Terminate, Treat or Tolerate. So, let's consider what Urquhart might have been able to do to manage the risks his division and his plan faced.

Stiffer than expected German resistance could not, in all probability, have been terminated; it could, however, have been treated. Urquhart, as shown, could have brought in more infantry than divisional support troops, thereby increasing his fighting strength and speed of movement.[11] We have also discussed that, perhaps more radically, Urquhart could have decided not to bring in 4th Parachute Brigade on the second day, or dropped it south of the river, this would have meant 1st Airlanding Brigade, his most powerful formation, would have been able to assist in the advance on the bridges on the first day.[12]

The possibility of poor weather would simply have to be tolerated as a risk. The only way to treat this issue would have been to have brought in the whole division on the first day. This was not possible given the air plan. Seasonal fog would indeed disrupt the third drop and effect resupply.[13] Any impact on CAS availability would also be a risk that would have to be similarly tolerated, although increasing the combat power by releasing 1st Airlanding Brigade from holding the landing and drop zones would have helped treat both risks to a limited extent.

We would argue the urban nature of the fighting was a risk Urquhart would have to tolerate in the main. There was not enough time to redress the lack of training in fighting in a built-up area; seven days was simply not long enough.[14] This issue could have been treated to a limited extent by the greater issue of maps of the town enabling shortcuts to be more easily identified.[15] There was, however, again probably not enough time to do this adequately. The issue might also have been mitigated by less focus on the main roads.

The reliance on the three main roads into Arnhem and the constraint this placed on tactical flexibility and manoeuvrability could have been treated by not pushing all three of 1st Parachute Brigade's battalions down separate

---

11 Harclerode, *Arnhem*, 162.
12 Frost, *A Drop Too Many*, 258.
13 Ritchie, *Arnhem*, 199.
14 Urquhart, *Arnhem*, 199.
15 Harvey, *Arnhem*, 50.

roads.[16] In addition, using the reconnaissance squadron in its intended role, rather than in a *coup de main* operation would have identified the lower road was more viable than the other two routes and this axis of advance could then have been used by all three battalions.[17] More maps and greater liaison with the Dutch would have helped treat this risk as shortcuts or alternative routes could have been identified more easily.[18]

Seizing the Heveadorp-Driel Ferry would have increased tactical flexibility, facilitating movement on both sides of the river.[19] Identifying the existence of the ferry and its importance would have helped treat the difficulties that restricted tactical movement. This is something the local Dutch resistance, and indeed Urquhart's Dutch liaison officers could have assisted with.[20] Further, if consulted more, the Dutch would have highlighted the importance of the heights. Holding this high ground, despite adding another objective for the division to accomplish, would have treated the problem of German ability to restrict the division's movement (and later prevent relief forces crossing the Rhine when XXX Corps did arrive).[21]

Finally, British radios. The failure in communications caused problems for Urquhart; it was one of the main reasons why he went forward to determine what was happening in Arnhem, then got stranded there during a critical period of the battle.[22] Functioning communications would have allowed Urquhart to redeploy his battalions to the southern route;[23] and also have helped coordinate CAS when it was available.[24] First Airborne Headquarters signallers were aware of the available equipment's limitations,[25] but, given the short timescale prior to the operation's launch, it was too late to do address this and so this risk would just have to be tolerated.

Likely, although none of the risks could have been terminated and two simply had to be tolerated, this analysis suggests that a number could have been treated if questions had been asked, and the external factors fully articulated at the time. This of course, is what the Outside-In View exercise is designed to do. The impact on the outcome of Market Garden and the part

---

16 Urquhart, *Arnhem*, 199.
17 Harclerode, *Arnhem*, 165.
18 Frost, *A Drop Too Many*, 200.
19 Powell, *The Devil's Birthday*, 241.
20 Middlebrook, *Arnhem 1944*, 444.
21 Mead, *General 'Boy'*, 145.
22 Frost, *A Drop Too Many*, 257.
23 Urquhart, *Arnhem*, 200.
24 Bennett, *A Magnificent Disaster*, 233.
25 Harclerode, *Arnhem*, 167.

of it that was the action at Arnhem in particular, that such an analysis would have provided, is an interesting point of speculation.

Conclusion

Using a Structured Analytical Technique, such as the Outside-In View exercise, might have helped Urquhart in his planning. Again, it is vital to note that here we have been working with the benefit of hindsight and so it is difficult to judge whether the factors we have discussed, could have been identified at the time, but, this is what the process is designed to do, and a number of the risks were known to some extent (in some quarters), at the time. Our analysis suggests some of the risks could have been treated. Again, these risk management treatments have been identified as suggestions by participants in the battle, such as Frost and Urquhart. Had they been put in place, these measures would most likely have produced greater combat power, enabled greater tactical flexibility and surely improved 1st Airborne's ability to fight its way through to and hold the Arnhem crossings.

A further note of caution. Ritchie warns of the problems of conducting counter-factual exercises. One of the issues that he points out is that, in the case of Arnhem, the counter-factual arguments do not sufficiently factor in the German perspective and likely reactions. For example, the Germans might have attacked Arnhem bridge in greater strength if 1st Airborne had been there in greater force. Similarly, a stronger attack on Arnhem bridge might conceivably have led Field Marshal Model to destroy Nijmegen bridge and thus halt XXX Corps' advance.[26] It is, therefore, impossible to determine whether conducting an Outside-In View exercise would have ensured success, but it would, surely, have helped. There are so many differing viewpoints and tantalising 'what ifs' surrounding Market Garden, it is hard to come to a definitive conclusion, but this is what we shall attempt to do in the final section of this book.

---

26 Ritchie, *Arnhem*, 187-190.

Withdrawal

Introduction

We have attempted in this book to examine Market Garden through the lenses of both psychology and military history. Psychologists tend to describe thoughts, feelings, actions and perhaps even predict behaviour; we have attempted to do this here. Psychologists tend not to judge (certainly according to Carl Rogers); however, it is perhaps permissible, as we end, to survey the judgements about Market Garden from a military historian's perspective. Opinion is, of course, divided on the subject.

Historical Assessment

Hibbert argues that Market Garden was the right option given the strategic circumstances and that it would have worked had Eisenhower provided full logistical support to the strategy. He argues more supplies would have allowed the two flanking British Corps to have provided more effective support, and XXX Corps would have advanced more quickly.[1] He points out that German officers interrogated after the war along with the examination of captured documents indicate agreement that if fully backed, the operation would have succeeded.[2] Clark agrees to some extent. His conclusion is that Market Garden was complex, 'audacious and risky',[3] but was 'a risk worth taking'.[4] Ritchie highlights benefits accruing from the operation, arguing that Market Garden stopped the Wehrmacht from establishing a defensive line along the Albert Canal and disrupted V2 attacks for a period. He accepts, however, that the operation was costly in terms of Allied lives and materiel and delayed the vital opening of the port at Antwerp.[5]

---

1 Hibbert, *The Battle of Arnhem*, 208.
2 Ibid., 201.
3 Clark, *Arnhem*, 100.
4 Ibid., 331.
5 Ritchie, *Arnhem*, 253-254.

Powell is amongst those who take a more negative view. He argues that the key strategic objectives were not achieved by the operation; the northern flank of the Siegfried Line was not outflanked, German Fifteenth Armee had not been encircled north of Antwerp and perhaps most importantly, the Ruhr was not taken.[6] He points out that even if it had been successful, it is unlikely the war would have ended by Christmas.[7] Harclerode goes one step further and quotes a former senior airborne officer who described the operation as 'the biggest cock-up of all time'. His view is that although it nearly succeeded, it was a failure, perhaps doomed from its moment of conception.[8] Bennett agrees that it could have succeeded but Allied planners ignored the danger signals, violated established principles of war and ignored hard won lessons.[9] He also points out that it may have succeeded at the operational level, by seizing a bridgehead over the Rhine but would have failed at the strategic level as any real follow-on exploitation would not have been possible.[10] Buckingham concurs,[11] as does Harvey.[12] Ritchie also agrees with this view, arguing that the Allies, in part because of Antwerp's unavailability, simply lacked the logistical wherewithal to exploit past Arnhem and into Germany.[13] This was Market Garden's strategic objective: to get Second Army over the Rhine, and so ultimately the operation was just too ambitious given the resources available. He argues the British Army was just not operationally capable of carrying out the ground part of the operation. He also concludes that the tactical objective to get a bridgehead over the Rhine was too late, whereas the strategic operation was too premature.[14] That final assessment, is probably a fitting way on which to end discussions on the operation, but what about the individuals?

Psychological Assessment

In this book we have tried to answer key questions about the three most senior British officers involved in the planning of Operation Market Garden. Why did Montgomery act so out of character in devising such a bold imaginative

---

6 Powell, *The Devil's Birthday*, 232.
7 Ibid., 244.
8 Harclerode, *Arnhem*, 11.
9 Bennett, *A Magnificent Disaster*, xii.
10 Ibid., 199.
11 Buckingham, *Arnhem 1944*, 235.
12 Harvey, *Arnhem*, 187.
13 Ritchie, *Arnhem*, 254.
14 Bennett, *A Magnificent Disaster*, 193-195.

plan? What drove Browning to push through an operation that was so high risk? How did Urquhart develop an overly simplistic and rigid plan? Hopefully we have answered these questions in the previous chapters, but to recap, lets summarise the psychological factors involved for each officer using the OODA Loop framework, working backwards from the point of failure in each case.

Montgomery

Montgomery committed several errors in the Act stage of the OODA Loop. These were driven by cognitive biases that were at play in his thinking during his planning for Operation Market Garden. His desperation to get what he saw as the required coherent strategy led him to adopt a narrow view of the options available to him. He grasped at the immediately available solution to the problem, using FAAA, then stuck too rigidly to his idea. This narrow focus meant that he failed to consider the pressing external factors that impinged on the operation, and he was too confidently focused on the internal planning considerations. This was his most significant error: taking a too narrow view of the options available to him to further prosecute the war in late August/ early September. He had lost his usual sense of realism and slipped from his habitual professional standards because of the frustrations he faced.

Montgomery's frustrations meant that he was, in terms of the coping strategy that he adopted in the preceding phase of the OODA Loop (Decide), experiencing Defensive Avoidance. His assessment was that there was clearly risk in the situation, the current dispersal of effort was threatening to prolong the war. German resistance was stiffening, his superior was failing to provide the necessary supplies, and his US counterparts (in his terms), were not being cooperative; there was, therefore from both a military and political perspective, no other viable solution available to him. In the face of this decisional conflict, he went into Bolstering mode and supported what he saw as the best option, Market Garden. In this mindset, he was impervious to the doubts and concerns of others and pushed on with an operation that was becoming increasingly unrealistic.

Focusing on the Orient stage of the OODA Loop, Montgomery's Bolstering behaviour was the outcome of his mindset at the time, in this instance, his Grip reaction. We have examined the evidence that suggests that he was, in Myers-Briggs terms, an ISTJ. In the grip of his inferior function, he moved away from his usual sense of realism and came up with a bold imaginative plan that by the time it was implemented, was impractical. His Grip reaction meant that he lacked his usual 'grip' (pun intended) on the conduct of the operation. This Grip reaction was triggered by several factors, including the looming deadline of the critical point whereby it would be too late to take advantage of the German dislocation, the uncertainty and confusion caused by a lack of a coherent strategy, and what he saw as Eisenhower's incompetence. The situational pressures he faced, added more stress and pressure; these were most evident at the Observe stage of the OODA Loop.

A couple of situational factors created additional pressures on top of the difficult task he faced. Ultimately, Eisenhower approved the Market Garden plan and directed Montgomery to ensure it was carried out. There was, therefore, explicit authority coming from the military chain of command. There was also considerable political pressure being applied and a great deal of interest in the operation at senior military levels. The interest shown by such eminent figures as Churchill, Brooke, and Marshall meant there was a great deal of implicit authority at play. The Scarcity Principle also applied further pressure. Stiffening German resistance meant that the opportunity provided by the Wehrmacht's dislocation would soon pass; if the Allies were to take advantage of the immediate situation, they needed to act quickly. Time was running out for Montgomery, and his frustration at this, was the main reason why he persisted with an operation that, by the time it was launched, was unsound. Commentators argue his persistence in continuing with Market Garden when it was too late means he should shoulder a good deal of the responsibility for its failure. He did much to push it through; and his collusion with his airborne advisor did much to influence Browning.

Browning

Browning also committed errors in the Act stage of the OODA Loop in that he was too personally invested in Market Garden and failed to challenge the viability of the operation. In terms of biases, he was too endowed in the wider airborne forces concept to see it not used in anger; Loss Aversion and missing out on the chance to go into action also drove him forward. Confirmation Bias also helped him to accept the downplaying of intelligence regarding II SS Panzer Corps. He was not critical enough and could or should have done more to push back on the plan.

His approach was determined by the problem coping strategy that he adopted in the Decide phase of the OODA Loop. Browning's perception that there was not another viable solution to a risky problem meant that he was in the Defensive Avoidance mindset, in particular Bolstering the best (or least-worst) option, Market Garden. In this state, his information-processing, as we have seen, was not as thorough and critical as it could have been. It can be argued that it even led him to actively 'manage' the situation; he may not have suppressed the available intelligence, as has been claimed, but it seems he did take steps to keep dissenters quiet.

Browning's adoption of the Defensive Avoidance (Bolstering) coping strategy was determined partly by the situation and partly by his personal filters, in his case motivation, that served as a driver at the Orient stage of the OODA Loop. His strong power motivation manifested itself in his ambitious streak. Market Garden was his last chance to get the operational command that he craved and wanted. He was, therefore, predisposed to the operation going ahead and thus prone to the pressures applied by the task and situational factors. These pressures exacted their influence at the Observe stage of the OODA Loop.

In terms of task pressures, he was given command of a difficult and complex operation. He was required to insert three and a half airborne divisions along a 65-mile corridor and seize several consecutive bridges, holding them, in the case of Arnhem, for two days. All of this needed to be put in place within seven days. If the operation succeeded, it was seen as shortening the war, if it failed, it would probably prolong it. He was under a lot of pressure to pull it off.

Added to this problem-induced pressure were several situational factors. He was given a lawful order, this meant he was required to be obedient to the explicit authority of the military command structure. The interest in the operation at senior levels meant that he was subject to a good deal of implicit authority as well. Another factor at play was Commitment and Consistency. Browning had invested substantial time and energy in developing and growing the British Airborne Forces. He was, quite naturally, a strong advocate for its use in the strategic role for which it had been created and was, therefore, very keen to see the payoff from this investment. Further pressure was applied by the Scarcity Principle. The general view that the war was ending, and the approaching autumn/winter meant the window of opportunity was rapidly closing; effectively this was his last chance. These task and situation pressures combined with his strong ambition ultimately led to a lapse in his usual professional standards when it came to Market Garden. He certainly did not benefit from Market Garden. Some would argue that there is an element of justice in this as he should share substantial blame for the failure of the operation. He did much to push it through; his support for Market Garden certainly had an impact on Urquhart.

Urquhart

Finally, let's summarise Urquhart's decision-making through the framework of the OODA Loop. Urquhart did commit errors in the Act stage in that he tackled the planning tasks facing him in too rigid and simplistic a manner. His planning was quite narrow in its focus and failed to identify and consider important points a broader, more imaginative approach might have raised. This then led him to make specific errors in his planning. He became anchored on the Comet concept of operations and failed to make sufficient adjustments for the changed requirements of Market Garden. He also suffered from a premature closure in his thinking due to the short timescales, subsequently failing to look at broader considerations.

His approach was determined by the problem coping or decision-making strategy that he adopted in the Decide phase of the OODA Loop. Urquhart's perception of the need to find a solution to a risky problem in a short timescale meant he adopted a coping strategy of Hypervigilance (before shifting towards Bolstering at the end). In this state, whilst actively engaged in the planning for the operation, his information-processing and decision-making activities as previously described were, however, rushed, unsystematic, incomplete and relatively simplistic in nature. This curtailed process precluded extensive information search and decision-making and led to a reliance on the more heuristic-based judgements and simplistic solutions described above.

Urquhart's adoption of the Hypervigilant coping strategy was driven partly by the circumstances of the situation and partly by his personal qualities (cognitive capacity) that he relied upon at the Orient stage of the OODA Loop. Urquhart's low to moderate Conceptual Complexity led him to rely on his previous (infantry and not airborne) experience to (inappropriately) make sense of the challenges he faced.

These challenges, both task-related and situational, were considerable and manifested themselves at the Observe stage of the OODA Loop. In terms of task pressures, he was asked to tackle a very difficult problem. In particular, the constrictive air plan for transporting his division created severe difficulties for him. Essentially, he would lose the element of surprise through being unable to effectively use *coup de main* parties and being forced to drop a long way from the bridges. With his division coming in over several days, he would also struggle to concentrate enough combat power to fight his way through to the bridges. His recognition and acceptance of these problems meant he was under a lot of pressure. Added to this problem-induced pressure were several situational influence factors.

The orders he received meant he was required to be obedient to the explicit authority of the military command structure. The interest in the operation at senior levels meant that he was subject to a good deal of implicit authority as well. Added to this, the lack of dissension about the plan meant that there would have been a lot of pressure for him to conform to the prevailing mood of optimism. The formation of the airborne forces in general and FAAA in particular meant there was a strong commitment to using 1st Airborne in the role for which it had trained; this pressure was only increased by the 16 operations that had been cancelled. More pressure was added due to the closing window of opportunity as the war appeared to be ending soon and his division thus threatened with disbandment. Urquhart was therefore given a difficult task and faced a lot of pressure to succeed. He was encouraged by Browning (whose own personal ambitions were clouding his judgement) whose enthusiasm for the operation meant that he supported Urquhart's plan and failed to provide any critical appraisal of it. Ultimately it is these task and situation pressures that caused him to act in the manner he did and therefore we would argue that he was ill served by the circumstances he faced. The final question, then, is could any of these issues or errors have been averted; would Urquhart, but also Montgomery and Browning been well served by thinking more critically about the situation?

Critical Thinking

Montgomery, Browning and Urquhart clearly committed errors. The main purpose of this book, however, is not just to describe what these were, but to use different psychological models to explore why they committed them. These models have been used retrospectively and descriptively in this book; but, because they define thinking and behaviour patterns, they could be incorporated into a Critical Thinking approach to support command decision-making in future operations.

The social influence pressures, obedience to authority, conformity and so on are well understood processes. These could be included in a situational awareness checklist to examine if a decision-maker has been subject to these pressures. The three motivational drivers can be assessed both in terms of the habitual motives that typically drive an individual, and more dynamically to establish which are being energised in a particular setting. Cognitive capacity can also be assessed to determine the challenges that an individual might face in terms of information-processing and decision-making when dealing with problem situations. Further dynamic decision support can be offered by examining the different components of the Decisional Conflict Model. For example, Structured Analytical Techniques could be used to critically evaluate which antecedent conditions are present in any situation. Observation of behaviour could further be used to monitor which of the coping strategies are being adopted. Checklists of structured problem-solving techniques, such as Brainstorming, could be used to encourage vigilant information-processing. Finally, the complexity of an individual's problem-solving can be monitored and processes implemented to encourage more complex thinking.

Checklists of the more common biases and judgement errors could also be used in conjunction with established Structured Analytical Techniques to quality assure plans. In this way, the psychological models discussed in this study could be used to structure a comprehensive Critical Thinking capability to support commanders in the future. It is interesting to speculate, whether the application of this Critical Thinking process would have mitigated the errors committed by our three decision-makers in 1944. For example, as demonstrated, Montgomery was overly focussed on getting a crossing over the Rhine. We used the Analysis of Competing Hypotheses in Chapter Five to examine the strategic situation. Another exercise, such as Issue Redefinition would have reduced the risk of him missing important external issues early in the planning process by examining the problem in different ways by rephrasing it, broadening or narrowing the focus or indeed reversing the question of whether he needed to get across the Rhine. This approach, if used early in the planning process, might have helped him reconsider his strategy.

The key issue for Browning was the viability of Market Garden. In Chapter Ten, we used the Cone of Plausibility exercise to examine the risk involved and the luck that would be needed for it to work. Again, we could have used other techniques. What if? analysis can help individuals consider how they are framing their problem and thus their perception of risk by assessing in a systematic way its potential gains and losses. In this way, Browning might have analysed the level of risk in the situation in a more robust manner. He was strongly committed to seeing Market Garden go ahead and failed to assess the situation as critically as he could have done. High Impact – Low Probability analysis involves identifying unlikely issues or events that might have a significant impact on a plan, it might have helped him consider more thoroughly the implications of the intelligence he was receiving.

Urquhart might have benefitted from a review of his plan for his division's role in Market Garden; in Chapter Fifteen we used the Outside-In View exercise to help with this. Using other analytical exercises such as the Structured Self-Critique technique might have enabled him to more thoroughly evaluate the quality of his information processing. This technique uses a structured process to identify sources of uncertainty and information gaps and might have led him and his planning team to identify key features they missed, such as the Heveadorp-Driel Ferry. He was also confident in his force composition based on his previous infantry experience and the belief airborne forces once landed on the ground were like regular infantry. An Assumptions Check exercise, where he could have examined the validity of this assumption might have led him to restructure his division's deployment. Another technique, Devil's Advocacy, might have also helped. This process involves challenging the confidence in a plan by arguing that it will fail and more importantly identifies possible reasons why. This process might have highlighted the difficulties of getting into Arnhem if his forces encountered stiffer than expected German resistance and narrow chokepoints on the routes in. This is all, of course, hypothetical and we can never be certain with counter-factual arguments, but it is interesting to consider what might have happened if Structured Analytical Techniques had been used in the planning for Operation Market Garden; but we think enough is enough now.

Final Word

This book has sought to explore and examine in detail the decision-making that manifested itself in the planning of Operation Market Garden in September 1944. We have examined the personal traits impacting Montgomery, Browning and Urquhart's judgements and which may have led to errors in their planning. As discussed at length in other works, errors were made. These errors, however, were made by individuals with different psychological characteristics that interacted in predictable ways with the external and internal pressures they faced. In conclusion, we have suggested that the errors committed by Montgomery, Browning and Urquhart were caused by the interaction between their psychological characteristics (personality, motivation and cognitive complexity), and the pressures placed upon them. Montgomery, at the strategic level, frustrated by a deteriorating situation and the inefficiency of others, was in the Grip of his inferior function, and constructed an impractical scheme to solve the problem. Browning, in command of the operation, under pressure to act and driven by his own ambition, did not do enough to challenge the plan. Urquhart, at the divisional level, faced with even stronger pressures to act and thinking too simplistically, devised a plan that was too rigid and narrowly focussed.

Ultimately, this book set out to refute 'bloody fool' theory and try to explain but not excuse the actions of the three officers concerned by examining the psychological factors at play: situational pressures, personality, motivation, cognitive complexity, coping strategy and biases. We are all obedient to authority or conforming to the group at times. We all have a Myers-Briggs type with an inferior Grip reaction that makes us act out of character. We are all, to a varying degree, driven by the Power motive that might not bring out the best-informed side of us. We all have different levels of ability to think in complex terms, sometimes we are too simplistic in our problem solving. We can all be too rushed in our thinking or simply resigned to making the best of a bad job. We all misapply our experience or fail to consider external factors in our planning. So, we will leave you with one final question; given all of this, do you think you would have acted differently or coped better? To end on a personal note, I don't think I would have.

BIBLIOGRAPHY

Archival materials

AHB, 1st Airborne Division Report on Operation Market, 10 January 1945.

LHCMA, Dempsey Papers, British 2nd Army Daily Intelligence Summary, No. 93, 5 September 1944.

LHCMA, Dempsey Papers, British 2nd Army Intelligence Summary, No. 101, 13 September 1944.

LHCMA, Dempsey Papers, British 2nd Army Intelligence Summary, No. 113, 25 September 1944, Part 2.

TNA, AIR 20/2333, 16th SS Panzer Grenadier & Reserve Battalion Report, Arnhem.

TNA: AIR 37/217: Information from Northern Group of Armies, Second Army and XXX Corps, as at 1100 hrs, 12th September 1944, by Lieutenant-Colonel A. Tasker, G-2, FAAA, 12 September 1944.

TNA, AIR 37/1214, 1st British Airborne Corps – Allied Airborne Operations in Holland Sept-Oct 1944.

TNA, AIR 37/1217, Operation Market, 1st Airborne Division Planning Intelligence Summary, No.2 dated 14th September 1944, prepared by G2(I), 1st Airborne Division, 14 September 1944.

TNA, AIR 37/1249, 21 Army Group, Operation Market Garden, 17-26 Sept 1944.

TNA, CAB 106/1133, official historian's notes of an interview with Lieutenant General Sir Frederick Browning, 7 October 1954.

TNA: DEFE 3/221, XL 9245, 6 September 1944.

TNA, WO 32/10899, Directorate of Staff Duties: Manpower.

TNA, WO 106/972, 1st Airborne Corps Operation Instruction No.1 Para 1.

TNA, WO 171/133, 21 Army Group Intelligence Review, No.160, 18 Sep 1944.

TNA, WO 171/341, 1st Airborne Division War Dairy, Planning Intelligence Summary 2 of 7, September 1944.

Bibliography

TNA, WO 171/366 1st Airborne Corps War Diary: Operation Market Instruction #1, 13 September 1944.

TNA, WO 171/393, 1 Airborne Div. Op Instr No.9, Confirmatory Notes on GOC's Verbal Orders, 1st Airborne Division War Diary, September-December 1944.

TNA, WO 171/393, 1 Airborne Division Planning Intelligence Summary No.1, HQ 1st Airborne Division War Diary, 1st-30th September 1944.

TNA: WO 171/393, 1st Airborne Division Planning Intelligence Summary No.2, 1st Airborne Division War Diary, September-December 1944.

WO 171/393, 1 Airborne Division Report on Operation Market, Part V, HQ 1st Airborne Division War Diary, 1st-30th September 1944.

TNA, WO 171/393, 1st Airborne Division War Diary, September-December 1944.

TNA, WO 171/393, 1st Airborne Division War Diary, September 1944, Annexure N, Operation Market, Story of 1 Parachute Brigade.

TNA, WO 171/393, 1 Para Bde OO No. 1, 1st Airborne Division War Diary, September – December 1944.

TNA, WO 205/313 21st Army Group Operation 'Market Garden' Plans and Instructions.

TNA, WO 205/313, Operation Instruction No. 2 – British Airborne Corps, 14th September 1944.

TNA, WO 205/313, Operation Orders from Brereton to Browning, 11th September 1944.

TNA, WO 205/873, Allied Airborne Operations in Holland, September to October 1944.

TNA, WO 205/874, IX Troop Carrier Command Report on Operation Market.

TNA, WO 208/3575, Brigadier E.T. Williams, 'Notes on the use [of Ultra],' 5 October 1945, 2.

TNA, WO 219/5137, 1 Parachute Brigade Intelligence Summary No.1, by Capt W.A. Taylor, 13 September 1944.

TNA, WO 219/1922, SHAEF Intelligence Summary No.23, 26th August 1944.

TNA, WO 219/4998, Operation Sixteen Outline Plan, 10 September 1944.

TNA: WO 285/3, Second Army Intelligence Summary, 6 September 1944.

TNA, WO 285/9, Dempsey Diary, 9 September 1944.

TNA, WO 285/9, Dempsey Papers, Personal War Diary as Commander 2nd Army, 'First 100 Days'.

TNA, WO 285/29, Dempsey to Ellis, 18 June 1962.

TNA, WO 285/29, Dempsey to Ellis, 7 July 1966.

CRCP 108/5, Cornelius Ryan Collection of World War II Papers, Mahn Center for Archives and Special Collections, Ohio University, Athens, Ohio.

Roy Urquhart, *Rising Eighty*, Urquhart Papers, Imperial War Museum, 102.

Official publications

Developments, Concepts & Doctrine Centre, *Land Operations* (Shrivenham, Army Doctrine Publication, AC 71940, 2017), 5-6.

Development, Concepts & Doctrine Centre, *Understanding & Decision Making* (Shrivenham: Joint Doctrine Publication 04, 2016), 68.

Journal articles

Brehmer, B. 'Dynamic Decision-Making: Human Control of Complex Systems', *Acta Psychologica*, 81, (1992), 211-241.

"Carbuncle". 'On an Excess of Bridges', *British Army Review*, No.108.

Folkman, S. 'Dynamics of a stressful encounter: Cognitive appraisal, coping, and encounter outcomes', *Journal of Personality & Social Psychology*, 50, 5, (1986), 992-1003.

Hammond, K. 'Judgement and decision making in dynamic tasks', *Information & Decision Technologies*, 14, (1988), 3-14.

Herek, G et al. 'Decision making during international crises. Is quality of process related to outcome? *Journal of Conflict Resolution*, 31, 2, (1987), 203-226.

Herman, M. 'Explaining foreign policy behaviour using personal characteristics of political leaders', *International Studies Quarterly*, 24, (1980), 7-46.

Houghton, D. 'Understanding Groupthink: The Case of Operation Market Garden', *Parameters*, 45, 3, (2015), 75-85.

Janis, I & Mann, L. 'Coping with Decisional Conflict', *American Scientist*, 64, (1976), 657-667.

Janis, I & Mann, L. 'Emergency Decision Making: A Theoretical Analysis of Responses to Disaster Warnings', *Journal of Human Stress*, 3(2), (1977), 35-48.

Keinan, G et al. 'Chunking and integration: Effects of stress on the structuring of information', *Cognition & Emotion*, 5, 2, (1987), 133-145.

Porter, C & Suedfeld, P. 'Integrative Complexity in the Correspondence of Literary Figures: Effects of Personal and Societal Stress', *Journal of Personality & Social Psychology*, 40, (1981), 321-330.

Raphael, T. 'Integrative complexity theory and forecasting international crises', *Journal of Conflict Resolution*, 26, 3, (1982), 423-450.
Suedfeld, P. 'Cognitive Managers and Their Critics', *Political Psychology*, 13, 3, (1992), 435-451.
Suedfeld, P & Bluck, S. 'Changes in integrative complexity prior to surprise attacks', *Journal of Conflict Resolution*, 32, 4, (1988), 626-635.
Suedfeld, P, Corteen, R & McCormick, C. 'The role of integrative complexity in military and leadership: Robert E Lee his opponents', *Journal of Applied Social Psychology*, 16, 6, (1986), 498-507.
Suedfeld, P & Rank, D. 'Revolutionary leaders: Long-term success as a function of changes in conceptual complexity' *Journal of Personality & Social Psychology*, 34, 2, (1976), 169-178.
Suedfeld, P & Tetlock, P. 'Integrative Complexity of Communications in International Crises', *Journal of Conflict Resolution*, 21, 1, (1977), 169-184.
Suedfeld, P, Tetlock, P & Ramirez, C. 'War, Peace and Integrative Complexity: UN Speeches on the Middle East Problem, 1947-1976', *Journal of Conflict Resolution*, 21, 3, (1977), 427-441.
Tetlock, P. 'Identifying victims of groupthink from public statements of decision makers', *Journal of Personality & Social Psychology*, 37, 8, (1979), 1314-1324.
Walker, S. 'Psychodynamic processes and framing effects in foreign policy decision making: Woodrow Wilson's operational code', *Political Psychology*, 16, 4, (1995), 697-717.
Winter, D. 'Leader appeal, leader performance, and motive profiles of leaders and followers: A study of American presidents and elections', *Journal of Personality and Social Psychology*, 52, (1987), 196-202.

Academic papers

Bradley, P. *Market Garden: Was Intelligence Responsible for the Failure?* (Maxwell Airforce Base, Alabama: Defense Technical Information Center, 2001).
Cirillo, R. *'The Market Garden Campaign: Allied Operational Command in Northwest Europe,1944'* (unpublished doctoral thesis, Cranfield University College of Defence Technology, 2002), 7-50.
Clemmesen, M. 'Combat Case History in Advanced Officer Development: Extracting What is Difficult to Apply', Baltic *Security & Defence Review*, 17, 2, (2014), 34-79

Coble, E. *Operation Market Garden: Case Study for Analyzing Senior Leader Responsibilities* (Carlise, PA: United States Army War College, 2009).

Dickson, J. '"But the Germans, General, the Germans, what about them?": The British Assessment of German Fighting Ability and Operation Market Garden, August to September 1944' (unpublished MA thesis, University of Buckingham, 2017), 128.

Hoyer, B. *Operation Market Garden: The Battle for Arnhem* (Maxwell Airforce Base, Alabama: Defense Technical Information Center, 2008).

Green, W. *Operation Market Garden* (Maxwell Air Force Base, Alabama: Defense Technical Information Center, 1984).

Greenacre, J. 'Assessing the Reasons for Failure: 1st British Airborne Division Signal Communications during Operation 'Market Garden"', *Defence Studies*, 4, 3, (2004), 283-308.

Jeffson, J. *Operation Market Garden: Ultra Intelligence Ignored* (Bolling Air Force Base, Washington DC: Joint Military Intelligence College, 1998).

Van Hook, G. *Tactical Victory Leading to Strategic Defeat: Historic Examples of Hidden Failures in Operational Art.* (Rhode Island: Naval War College, 1993).

Books

Adair, J. *Action-Centred Leadership*. London: Gower, 1979.

Baynes, J. *Urquhart of Arnhem*. London: Brassey's, 1993.

Beevor, A., *Ardennes 1944: Hitler's Last Gamble*. (London: Penguin, 2015.

Beevor, A. *Arnhem: The Battle for the Bridges*. London: Penguin, 2018.

Bennett, D. *A Magnificent Disaster: The Failure of Market Garden, The Arnhem Operation, September 1944.* Newbury: Casemate, 2008.

Brereton, L. *The Brereton Diaries – The War in the Air in the Pacific, Middle East and Europe 3 October 1941 – 8 May 1945*. William Morrow: New York, 1946.

Briggs Myers, I. *Introduction to Type*. Oxford: OPP, 1994.

Buckingham, W. *Arnhem 1944*. Stroud: Tempus, 2015.

Buckley, J. *Monty's Men: The British Army and the Liberation of Europe*. London: Yale University Press, 2014.

Buckley, J & Preston-Hough, P, *Operation Market Garden: The Campaign for the Low Countries, Autumn 1944: Seventy Years On*. Solihull: Helion, 2016.

Carver, M. *The Seven Ages of the British Army*. London: Harper Collins, 1986.

Chabris, C & Simons, D. *The Invisible Gorilla*. London: Harper Collins, 2011.

Cialdini, R. *Influence: Science & Practice*. Boston: Allyn & Bacon, 2001.

Cirillo, R. 'Market Garden and the Strategy of the Northwest Europe Campaign'. *Operation Market Garden: The Campaign for the Low Countries, Autumn 1944: Seventy Years On*, edited by Buckley & Preston-Hough, Solihull: Helion, 2016.

Clark, L. *Arnhem: Jumping the Rhine 1944 & 1945*. London: Headline, 2008.

Clark, R & Mitchell, W. *Deception: Counterdeception and Counterintelligence*. London: Sage, 2019.

Cohen, E & Gooch, J. *Military Misfortunes: The Anatomy of Failure in War*. New York: Free Press, 1990.

Colville, J. *The Fringes of Power – Downing Street Diaries 1939-1955*. London: Wedenfeld & Nicolson, 2004.

Didden, J. 'A Week Too Late?' *Operation Market Garden: The Campaign for the Low Countries, Autumn 1944: Seventy Years On*, edited by John Buckley & Peter Preston-Hough, Solihull: Helion, 2016.

Dixon, N. *On the Psychology of Military Incompetence*. London: Futura, 1976.

Dover, V. *The Sky Generals*. London: Cassell, 1981.

Fredholm, L. 'Decision making in firefighting and rescue operations'. *Sitting in the Hot Seat*, edited by Rhona Flin, London: Wiley, 1996.

Frost, J. *A Drop Too Many*. London: Pen & Sword, 1994.

Gavin, J. *On to Berlin: Battles of an Airborne Commander 1943-1946*. New York: Viking Press, 1978.

Gigerenzer, G & Todd, P. *Simple Heuristics That Make Us Smart*. Oxford: Oxford University Press, 1999.

Gilchrist, R. *Malta Strikes Back: The Story of 231 Brigade*. Uckfield: The Naval & Military Press, 1945.

Hamilton, N. *Monty: Master of the Battlefield 1942-1944*. London: Hamish Hamilton, 1983.

Hamilton, N. *Monty: The Field Marshal 1944-1976*. London: Hamish Hamilton, 1986.

Hamilton, N. *Monty: The Battles of Field Marshal Bernard Montgomery*. London: Hodder & Stoughton, 1994.

Hamilton, N. *The Full Monty: Montgomery of Alamein 1887-1942*. London: Penguin, 2001.

Harclerode, P. *Arnhem: A Tragedy of Errors*. London: Caxton, 1994.

Hardman, D. *Judgement and Decision Making: Psychological Perspectives*. Chichester: Blackwell, 2009.

Harvey, A. *Arnhem*. London: Cassell, 2001.

Henniker, M. *An Image of War*. London: Leo Cooper, 1987.

Heuer, R & Pherson, R. *Structured Analytical Techniques for Intelligence Analysis.* Los Angeles: Sage, 2015.

Hewstone, M et al. *An Introduction to Social Psychology.* Chichester: Wiley, 2015.

Hibbert, C. *The Battle of Arnhem.* London: Batsford, 1962.

Hirsh, S & Kummerow, J. *Introduction to Type in Organisations,* Oxford: OPP, 1994.

Holmes, R. *Battlefields of the Second World War.* London: BBC, 2001.

Irving, D. *The War Between the Generals.* London: Penguin Books: 1981.

Janis, I. *Groupthink.* Boston: Wadsworth, 1982.

Janis, I & Mann, L. *Decision Making: A Psychological Analysis of Conflict, Choice & Commitment.* New York: Free Press, 1979.

Jaques, E. *Requisite Organisation.* Arlington: Cason-Hall, 1998.

Kahneman, D. *Thinking, Fast and Slow.* London: Penguin, 2011.

Keegan, J. *The Mask of Command: A Study of Generalship.* London: Random House, 1987.

Kershaw, R. *It Never Snows in September.* London: Ian Allan, 1990.

Koestner, R & McClelland, D. 'The affiliation motive'. *Motivation and personality – Handbook of thematic content analysis,* edited by Charles Smith, New York: Cambridge University Press, 1992.

Maio G & Haddock, H. *The Psychology of Attitudes & Attitude Change.* London: Sage, 2015.

McClelland, D. 'How do self-attributed and implicit motives differ?' *Motivation and personality – Handbook of thematic content analysis,* edited by Charles Smith, New York: Cambridge University Press, 1992.

McClelland, D & Koestner, R. 'The Achievement Motive'. *Motivation and personality – Handbook of thematic content analysis,* edited by Charles Smith, New York: Cambridge University Press, 1992.

Mead, R. *General 'Boy': The Life of Lieutenant-General Sir Frederick Browning.* London: Pen & Sword, 2010.

Middlebrook, R. *Arnhem 1944: The Airborne Battle.* London: Penguin, 1994.

Montgomery, B. *The Memoirs of Field Marshal Montgomery.* London: Pen & Sword, 1958.

Nicholl, T & Rennel, T. *Arnhem: The Battle for Survival.* London: Penguin, 2011.

Otway, T. *Airborne Forces.* War Office official monograph, 1951.

Peaty, J. 'Operation Market Garden: The Manpower Factor'. *Operation Market Garden: The Campaign for the Low Countries, Autumn 1944: Seventy Years On,* edited by John Buckley & Peter Preston-Hough, Solihull: Helion, 2016.

Plous, S. *The Psychology of Judgment and Decision Making*. New York: McGraw-Hill, 1993.
Powell, G. *The Devil's Birthday: The Bridges to Arnhem 1944*. London: Papermac, 1985.
Quenk, N. *In the Grip: Understanding Type, Stress and the Inferior Function*. Oxford: CPP, 2000.
Ritchie, S. *Arnhem: Myth and Reality, Airborne Warfare, Air Power and the Failure of Operation Market Garden*. London: Robert Hale, 2011.
Ritchie, S. 'Learning to Lose? Airborne Lessons and the Failure of Operation Market Garden'. *Operation Market Garden: The Campaign for the Low Countries, Autumn 1944: Seventy Years On*, edited by John Buckley & Peter Preston-Hough, Solihull: Helion, 2016.
Ryan, C. *A Bridge Too Far*. London: Hodder & Stoughton, 1974.
Shortland, N, Alison, L & Moran, J. *Conflict: How Soldiers Make Impossible Decisions*. Oxford: Oxford University Press, 2019.
Sosabowski, S. *Freely I Served: The Memoir of the Commander – 1st Polish Independent Parachute Brigade 1941-1944*. London: Pen & Sword, 2013.
Suedfeld, P et al. 'Conceptual/integrative complexity'. *Motivation and personality – Handbook of thematic content analysis*, edited by Charles Smith, New York: Cambridge University Press, 1992.
von Clausewitz, C. *On War*. Princeton: Princeton University Press: 1984.
Urquhart, R. *Arnhem*: London: Pen & Sword, 1958.
Veroff, J. 'Power motivation'. *Motivation and personality – Handbook of thematic content analysis*, edited by Charles Smith, New York: Cambridge University Press, 1992.
Warren, J. *Airborne Operations in World War II, European Theater* (USAF Historical Division, Research Studies Institute, Air University, 1956.
Winter, D. 'Content analysis of archival materials, personal documents, and everyday verbal protocols'. *Motivation and personality – Handbook of thematic content analysis*, edited by Charles Smith, New York: Cambridge University Press, 1992.

INDEX

1st Airborne Division, 135
   ability to generate combat power, 136
   headquarters, 214
   inept planning, 173, 193
   lack of maps, 97, 123, 203, 213–14
   radios, 33, 212
   Reconnaissance Squadron, 183, 193, 200–201, 203, 214
   threat of disbandment, 161–62, 177, 222
1st Airlanding Brigade, 136, 181, 192, 198, 213
1st Airborne Corps, 174, 181–82, 226–27
1st Parachute Brigade, 98, 118, 133, 136, 150, 155, 158, 181–83, 192–93, 195–96, 198, 201, 213, 227
2nd Parachute Battalion, 133, 155, 195, 201

Achievement Motive, 103–7, 232
Affiliation Motive, 103–7, 232
airborne
   ability of, 141, 143, 145
   concept, 98, 111, 115, 173
   doctrine, 124, 146, 182, 196
   fighting power, 148–49
   forces, 76–77, 80, 95–96, 98–99, 101–2, 108, 110–13, 123, 129, 131, 134–38, 141, 143, 145, 162, 182, 195, 206, 222, 232
   frustration, 157
   limitations, 131
airborne lessons, 7, 233
air lift, 121, 123–24, 180, 198
   plan, 6, 70, 121–22, 124–27, 137–39, 159, 174, 180, 182, 190, 213
air planners, 3, 16, 70, 122, 188, 205
Analysis of Competing Hypotheses, 81–83, 223
Anchoring Effect, 76–77, 199
Antwerp, 11, 84–85, 87
armed jeeps, misuse, 200
Arnhem, xiii–xv, xvii, 2–7, 10, 12–22, 24–27, 29, 37, 41, 43–50, 52, 56–59, 62–66, 69–70, 72–82, 84–87, 94–101, 109–13, 115–31, 133–34, 136–38, 141–42, 145–46, 148–51, 153–60, 163, 170–84, 186–88, 191–208, 211–17, 220, 224, 226, 230–33
   bridge, 120, 142, 145, 148–49, 196, 200, 215
   roads, xiii, 16–17, 19, 63, 77, 130, 141, 149, 150, 181, 192, 197, 199, 201–3, 212, 214, 224
   selection of, 16
   terrain, 15, 141, 181–82, 203, 211–12
   urban nature of fighting, 182, 203–4, 211–13

Index

Authority Principle, 23–24, 95, 101, 155–56, 173, 183, 222, 225
   explicit, 95, 97, 162, 219–20, 222
   implicit, 24, 95, 97, 101, 162, 219–20, 222
   lawful orders, 95, 97, 101, 154, 220
Availability Heuristic, 75, 201
Bad Case Scenario, 143, 147
Bedell-Smith, Walter, 40, 52, 65, 78–79
Belchem, David, 16
Best Case Scenario, 147, 149–51
Bolstering, 60, 65, 67, 114–27, 189–90, 220–21
Bradley, Omar, 17–19, 43–44, 46, 48–49, 86–87, 229
Brereton, Lewis, 95, 115
bridges, xiii, 2, 6, 16, 20–21, 27–28, 43–45, 48, 50, 52, 58–59, 62, 68, 77–79, 92, 119–20, 122–23, 126–27, 129–30, 141–42, 146–47, 148–49, 154, 157, 161, 177, 180–82, 184, 187–88, 190, 192–93, 196–201, 203–4, 208, 212–13, 222, 228, 230, 233
   destroyed, 145, 148
   key, 148–49
British Army, 13, 21, 27, 64, 93, 111, 155, 161, 194, 198, 206, 217, 230
   military practice, 198
British economy and manpower situation, 6, 13, 26, 62, 66, 75, 85, 88, 99, 161, 177, 226, 232
broad front strategy, 10, 17, 19–20, 24, 43, 74, 86, 88
Brooke, Alan, xv, 27, 44, 46–48, 98, 219
Browning, Frederick, xiv–xvi, xviii, 3, 5, 7, 20, 69–70, 76, 90–98, 100–102, 107–22, 125–28, 130–31, 133–36, 138–40, 150–51, 153–55, 158, 160, 173, 176–77, 181, 187–88, 193–94, 197, 200, 205, 207, 218–20, 222–23, 225–27, 232
   ambition, 107, 111, 127, 135
   appointment to command blocked, 96
   background, 92
   biases, 132
   Commander Para-Troops and Airborne Division, 108
   concerns, 125
   decision-making, 114–27
   disciplinarian, 108
   disliked, 96
   errors, 128
   FAAA, 140
   father of the airborne, 98, 100
   inexperience, 131–32
   manipulation, 110
   military career, 139
   motivation, 8, 101–3, 105, 107, 109, 111, 113
   pressures, 92–101
   prestige, 105, 109–10
   questioned, 187
   risk perception, 116
   staff ridiculing, 136
   Urquhart, xvi, 7, 205, 222, 224–25

Canadian First Army, 46
cancelled operations, impact of, 97, 160, 177, 183, 186, 207
casualties, xiii, 13, 35, 85, 88, 119, 148, 180
Chatterton, George, 200
Chief of the Imperial General Staff (CIGS), xiv, 44, 46, 98
Chiefs of Staff, British, 24
cognitive biases, xviii, 1–2, 7–8, 70–71, 73–74, 76, 80–82, 127, 132,

135, 146, 191, 193–94, 205, 208, 210, 218–19, 225
cognitive complexity, 34, 163–71, 173, 223
    capacity, 7–8, 163, 166, 170, 172, 174, 221, 223
    high levels, 168
    state level, 166
    continuum, 164
Cognitive Dissonance, 68–69, 71, 75, 77–79, 81, 132–33
combat power, 136, 143, 145, 148–49, 180–81, 198–200, 212–13, 215, 222
Comet, 16, 20–21, 25, 66, 76–77, 114, 117, 123, 125–26, 129, 133, 135, 157–59, 176–77, 186, 198, 207
    cancelled, 14, 58–59
    expansion to Market Garden, 66
    plan, 77, 116, 124, 126, 186, 200
    planning for, 117, 125
Conceptual Complexity, 165–66, 169–71, 229
    high levels of, 166, 170
    and Integrative Complexity, 166–67, 233
    and Mode, 170
Cone of Plausibility exercise, 140, 142–43, 223
Confirmation Bias, 132, 134, 197, 219
Conformity Principle, 95, 153, 155–57, 159, 161, 183, 222
Consistency Principle, xvi, 92–101, 134, 159–60, 183, 220
coping
    behaviours, 65, 125
    mechanisms, xviii, 78, 189
    problem-focused, 61
    problem-focussed, 61
    strategy, xviii, 53, 113–14, 126, 175, 218, 223, 225
coping pattern, 53–55, 167, 189–90

critical thinking, 82, 140, 222

Decisional Conflict Model (DCM) 53–56, 66, 73–74, 78, 114–15, 120, 124–25, 167, 185, 187, 190, 223
    coping patterns, 61, 176
    Defensive Avoidance, 60–61, 64–65, 125, 189, 218, 220
    Hypervigilance, 175–91, 203, 221
    model, 60, 180, 184–85, 187–88
    process, 59, 64
    Procrastination, 61, 189
    Scapegoating, xiv, 60, 65, 125, 189
    Unconflicted Adherence, 54–55
    Unconflicted Change, 55
    Vigilance, 53–54, 167, 207
D-Day airborne experience, 138
D-Day airborne operations, 163, 195
decision-making, xvi–xvii, 1–2, 4–5, 7, 10, 29–30, 34–35, 53–55, 70, 107, 150–51, 164, 167, 187, 205–6, 221–24
Dempsey, Miles, 57, 59, 70, 109, 125, 130, 170, 173, 193, 227
    Browning, 70, 125
    Montgomery, 57, 59
Dixon, Norman, 1
Down, Eric, 98, 110, 155
Dutch army staff appreciations, 17
Dutch liaison officers, 202, 204
Dutch resistance, 57, 117, 204
Dutch underground reports, 57, 65, 75, 78, 176

Eindhoven, 123, 186
Eisenhower, Dwight, 15, 17–22, 24–25, 27, 43–48, 57, 59, 68–69, 76, 84–87, 96–97, 99–100, 122, 150, 154, 216, 219
    Montgomery, 44, 154
    persuaded, xv

Index

strategy, 21
Supreme Headquarters Allied Expeditionary Forces, 11
emotions, 35, 61
Endowment Effect, 134–35
Exposure Effect, 194, 208

First Allied Airborne Army (FAAA), xiv, xvi, 14, 16, 19–21, 24, 49, 66, 75–76, 81, 84–85, 88, 95–101, 115, 135, 140, 153, 160, 162, 182, 205, 218, 222, 226
   Chief of intelligence, 66
   command, 95
flak concentrations, 16
flanking British Corps, 63, 216
Framing Effect, 73–74, 107, 194, 229
   Montgomery and Browning, 194
Frost, John, 5, 111, 118, 133, 155, 173–74, 177, 193, 198–200, 204, 211, 213–15, 231
   Urquhart, 215

Gale, Richard, 127
Garden, 221
   forces, 141
Gavin, James, 80, 96, 112, 127, 129–31, 142, 178–79, 196, 198, 200, 231
German
   armour, 57, 73, 115–18
   armour refitting, 118
   attacks, 40, 145, 148
   battlegroups, 179
   counter-intelligence service, 57
   defences, 25, 58, 82
   dislocation, 84, 219
   Fifteenth Armee, 217
   fighter gruppens and flak concentrations, 16
   forces, 58, 62, 65, 79, 84, 141, 143, 149, 159, 176–77, 180, 194, 206
   High Command, 13, 57
   reaction, 80, 145–46, 148–49, 157
   resistance, 6, 13, 45, 58–59, 63, 66, 69, 72, 77, 80–82, 116, 118–19, 127, 133, 140, 145–47, 148–49, 159, 161, 178, 197–98, 200, 212, 218
   troops, 182, 190
   weakened Panzer Corps, 56
German forces
   disorganised resistance, 129
   expected resistance, 213, 224
   growing resistance, 66
   reaction, 79, 143
   reorganisation, 25, 59
   stiff resistance, 13, 179
Goal Direction, 71–72
Good Case Scenario, 143, 147–51
Groesbeek Heights, 130–31, 142
ground corridor, xiii, 181
   single lane road, 145, 148
Groupthink, 3, 53, 205–9, 229, 232
   symptoms, 3, 207–8
Guards Armoured Division, 141
Guingand, Francis, 16, 18, 63, 81

Hackett, Shan, 119, 158, 177, 180, 193, 205, 207–8, 210
Heveadorp-Driel Ferry, 203–5, 211–12, 214, 224
Hibbert, Tony, 117, 119
Hopkinson, George, 98, 112

II SS Panzer Corps, 56, 58, 65, 68, 78, 80, 116, 119–20, 176, 197–98, 219
   regrouping, 140
impulsiveness, 48
Inattentional Blindness, 72

information processing, 29, 53–55, 70, 164, 170, 185, 190, 209, 220–21, 223–24
Integrative Complexity, 166–70, 193, 228–29
  high levels of, 167–68
  low levels of, 166, 168
intelligence, 3, 41, 52, 56, 65–66, 78–79, 96, 115, 118–19, 132–34, 139–40, 146, 176, 179, 219, 224
  challenging, 65, 67, 75
  limited Photo Reconnaissance, 134
  Market Garden, 132
  photographs, 117, 119
  picture, 78, 115, 141, 176, 179, 190, 197
  reports, 58, 66, 116, 132, 159
  summaries, xvii, 133–34
  suppressed, 115, 134

jeeps, armed, 183, 189, 193, 200–201
Judgement, and decision making, 70, 79–80, 135, 164, 210, 228, 231
judgement, heuristic-based, 190, 221

landing zones, 6, 68, 77, 121–22, 124–26, 129–30, 137–40, 142, 145–47, 148–49, 157–58, 181–84, 187–88, 190, 192, 196–98, 201, 208, 212–13
Lathbury, Gerald, 98, 110, 118–19, 133, 155, 192, 195, 200–201
Lessons, 7, 68, 136–38, 217
  correct, 136, 138
  wrong, 137
Linnet, 70, 122–23
logistics, 11–13, 15, 19–20, 22, 26, 44, 62, 64, 66, 85–89, 99, 145, 148, 172, 174, 184, 190, 216–18

Loss Aversion, 135, 219
losses, 48, 50, 74, 142, 194, 212, 223
  potential, 54, 73–74, 135–36, 194

Mainline Scenario, 143–45
Market, xiv–xv, 62, 77, 118, 121, 129, 177, 180, 195–97, 220, 226–27
Market Garden, xiii–xv, xvii–xviii, 1–8, 10, 12–14, 17–18, 20–21, 24–29, 41, 46, 49–52, 57, 59, 62–70, 72, 74, 76–81, 90, 92–94, 97, 99–102, 107, 109–13, 115–17, 119–29, 131–36, 138–39, 142–44, 150–51, 153, 157, 159, 161, 163, 170–71, 173–78, 180–82, 186–88, 191, 194, 196, 198, 205–8, 211, 214, 216–21, 223–24, 226, 228–33
  changed requirements, 221
  command, 95–96
  concept flawed, 129
  failure, 80, 138, 230
  feasibility, 59–61, 63, 65, 121, 124–25, 132, 140–49, 173, 180
  flaws, 69
  gamble, 50–51, 59, 65, 67, 73–74, 90, 147
  launch of, 52, 93
  objectives, 6, 22, 48, 68, 77, 84, 124–25, 129–30, 138, 146–47, 148–49, 157, 181, 183, 192, 196, 198–99, 208
  operational concept, xv
  outcome, 142, 214
  plan, 12, 21, 29, 100, 111, 180, 184, 186, 227
  planning for, 129, 173, 188, 208
  responsibility for failure, 69
  rushed planning, xvii, 202
  salient, 64, 103, 141, 145, 148–49
  short timescales, 3, 26, 95, 122, 208, 214, 221

Index

strategic defeat, 3, 230
viability, 64, 223
Marshall, George, 24, 219
Milgram Stanley, 23–24, 157
Mode, 169–71, 173–74, 199
Model, Walther, 57, 215
Modes, 169
Montgomery, Bernard, xiii–xvi,
  xviii, 3, 5, 7–10, 12–29, 31–37,
  39–52, 56–59, 61–72, 74–81, 84, 86,
  95, 97, 100, 109, 111, 114, 120, 122,
  125–26, 128, 130–32, 139, 150, 154,
  170, 193–94, 205, 218–19, 222–23,
  225, 231–32
  behaviour in relation to Market
    Garden, 49
  biases, 132, 201
  Browning, 128, 193–94, 222
  character, 39
  childhood, 38
  chivvying Eisenhower, 45
  coping strategy, 52
  decision-making, 52–67
  deliberate lack of consultation, 70
  difficult personality, 26–29, 35
  disliked, 29
  eccentricities and pretensions, 28
  ego, 29
  external considerations, 81
  frustrations, 18, 46, 218
  goal-direction, 72, 74
  Grip, 43
  Grip behaviours, 46
  Grip reaction, 43–44
  Introversion, 31, 38
  Judging, 37
  Land Forces Commander, 15, 18,
    45–46, 51
  lapses, 71
  on Market Garden, 120
  megalomania, 29
  memoirs, 18
  mindset, 52
  obedience to authority, 24
  penchant for meddling, 41
  personality, 8, 27–51
  plan, 68–81
  pressures, 10–25
  problems, 62, 76
  promotion to Field Marshal, 45
  risk calculus, 56, 73
  Sensing, 32
  Thinking, 35, 74, 77, 81
  Type, 39
  Urquhart, 111
  vanity and ambition, 29
motivation, xvi, 2, 94, 102–3, 107,
  166, 220, 225
motivational drivers, xviii, 7, 102–3,
  105, 223
motivation and personality, 103–5,
  166, 232–33
Myers-Briggs
  dominant functions, 41–42
  Extraversion-Introversion, 30–31,
    38
  Grip, 27–51, 218–19, 225, 233
  Grip reaction, 42–43, 50–51, 53,
    56, 74, 218–19
  Grip reaction for ISTJs, 45
  inferior function, 41–43, 46, 48,
    50–51, 218, 225, 233
  iNtuition, 32, 38
  ISTJ, 38–39, 42–43, 45, 48, 50, 218
  ISTJ Grip behaviours, 46
  Judging-Perceiving, 36–38
  model, 30, 34, 36, 38, 41–42, 50
  Sensing, 32–34, 37–38, 42
  Sensing-iNtuition, 32, 41
  Thinking-Feeling, 34, 41
Myers–Briggs Type Indicator
  (MBTI), 29

narrow salient, 141, 143, 145, 148–49
Nijmegen, xiv, 50, 57, 80, 123, 126, 128–30, 136, 186, 191, 196, 198
Nijmegen bridge, 6, 64, 130, 215
northern thrust, 15–16, 18–21, 24, 43, 46–47, 56–57, 59, 61–62, 65–66, 72, 74–76, 81, 86, 90, 100
Northwest Europe Campaign, 10, 18, 59, 69, 121, 125, 182, 230

OODA Loop, 8, 26, 51, 66–67, 102, 113–14, 128, 154, 162, 174–75, 218–21
Optimism Bias, 79–80, 195, 208
Outside-In View exercise, 214–15, 224

Panzer Divisions, 116, 190
    battle-scarred, 119
    depleted, 57, 118, 179
Patton, George, 12, 15, 17, 19, 22, 43–44, 74
personality, xvi, xviii, 5, 28–30, 35, 38–40, 51, 94, 103–6, 166, 225, 229, 232–33
Photo Reconnaissance (PR), 116
planning, xiii, xv–xviii, 1–3, 5, 7–8, 49–50, 69, 71, 79–81, 96, 98, 109, 114, 117, 122, 125, 129, 137–38, 143, 153, 159, 164, 170, 173, 175–76, 180, 184, 186, 188, 190–91, 201–2, 204, 208, 211, 215, 217–18, 221, 224–25
    errors, 129
    process, xvii, 70, 122, 187, 189–90, 204, 208, 211, 223
    staff, 49, 202, 206
Planning Fallacy, 80–81, 204–5, 208, 210
Polish Independent Parachute Brigade, 5, 99, 233

political leaders, 103, 106, 228
politics, 16–17, 95
Power and Affiliation motivation, 106
Power Motive, 103, 105–7, 109–11, 113, 225, 233
pragmatism, 33
Premature Closure, 53, 70, 203, 221
pressures, xvi, xviii, 14, 22, 24–26, 42, 44, 51, 53, 69, 72, 74, 34, 86–89, 93, 95, 99–100, 114, 134, 154, 156, 159, 161–63, 176, 186, 219–20, 222–23, 225
    group, 156–57, 183
    problem-induced, 220, 222
problem coping strategy, 219
psychological conflict, 51, 53, 56

RAF Planners, 121, 197
Requisite Complexity, 165
Rhine, 10, 15–17, 20–21, 24, 62, 68, 128, 131, 139, 145, 150–51, 191, 203–4, 212, 214, 217, 223, 231
Ridgway, Matthew, xiv, 96–97, 112, 115
risk, xiv, xvi, xviii, 1–2, 4, 6, 8, 12–80, 84, 86, 88, 90, 94, 96, 98, 100, 104–13, 115–16, 118–27, 130–32, 134, 136, 138, 140–215, 218, 220, 222–24, 228, 230, 232, 236, 238
    assessment, 56, 178, 194
    management, 211, 213
Royal Air Force (RAF), 13, 20, 98–99, 122, 180, 197, 200
Ruhr industrial region, 15–16, 20, 22, 46, 48, 57, 86–87, 217

Saar industrial region, 15, 22, 48, 86
Scarcity Principle, 23, 25, 74, 95, 100–101, 160, 183, 219–20
Scheldt Estuary, 11–12, 14, 84, 87

Second Army, xiv, 10–12, 14, 16, 41, 52, 57–58, 62, 65–66, 77, 82, 133–34, 158, 161, 217, 226
Selective Perception, 72
Sense of Invulnerability, 207
Siegfried Line, 11, 14–16, 72, 217
situational pressures, xviii, 22, 26, 29, 72, 74, 95, 101, 139, 154, 162, 219–20, 222, 225
social influence factors, 22, 134, 154, 162
Social Proof, 23, 156
SOE (Special Operations Executive), 57
Sosabowski, Stanislaw, 76, 114, 119–20, 134, 139, 153, 156, 158–59, 173, 178, 197, 205, 207, 210, 233
southern thrust, 43–44, 86
SS Panzer Divisions, 52, 58, 119
strategic bombing campaign, 87–88, 98
strategic situation, xv, 4, 10, 18, 56, 223
  best strategy, 14, 78
  debate, 74, 90
  options, 87
  thrusts, 15, 20, 44, 48, 62, 74, 86–89
Stratified Systems Theory, 171–72, 199
stress, 36, 39, 41–42, 47, 61, 70, 206, 219, 228, 233
Strong, Kenneth, 52
Structured Analytical Techniques, 81–82, 139–40, 209–10, 215, 223–24
Sunk Cost Effect, 193, 208
Supreme Headquarters Allied Expeditionary Forces (SHAEF), 11, 14, 49, 52, 56, 59, 79, 89, 99
surprise, loss of, 144, 197

Task Complexity, 165
time pressure, perceived, 25, 160
Twelfth Army Group, 10
Twenty-First Army Group, xv, 10, 13, 16, 18–19, 66, 70, 100, 120, 126, 150, 179

Ultra
  decrypts, 52, 56–57, 78, 176
  intelligence ignored, 2, 230
Urquhart, Brian, 29, 94, 109, 116, 119, 126, 129, 133–34, 139, 177
Urquhart Roy, xiii–xiv, xvi–xviii, 3, 5, 7–8, 63, 80, 109, 111, 114, 116–18, 125, 134, 136, 139, 142, 150–51, 152–56, 158–63, 165, 170–215, 218, 221–22, 224–25, 233
  character, 156, 172
  cognitive capacity, 163–73
  confidence in plan, 178
  coping strategy, 174
  decision-making, 175–89
  doubts about operation, 184
  force composition, 198, 224
  frustration, 157
  information processing, 172
  issues, 183
  lack of airborne experience, 3, 172, 174, 199
  lack of airborne inexperience, 155
  level of cognitive capacity, 174
  optimism, 178
  plan, 191–209
  planning ability, 182
  planning constraints, 6
  planning timeframe, 186
  pressures, 159
  problems, 191
  protests, 126
  risk calculus, 176
  sense of authority, 155

situation assessment, 176, 186
state of mind, 175
subordinates, 155
suicide mission, 175, 187, 189
US 82nd Airborne Division, 142
US 101st Airborne Division, 142
US First Army, 16, 19
US IX Troop Carrier Command, 123

V2 attacks, 24
victory euphoria, 12, 14, 25, 45, 79, 82, 100, 120, 176, 183

Walcheren Island, 84
weather, 68, 80–81, 142–43, 145, 150–51, 161, 211–13
Wehrmacht, 6, 11–15, 18, 58, 62, 81, 84, 133, 160, 179, 197, 216
Wesel, 16, 19–21, 57, 59, 72, 125
Arnhem, 19–20
Westerbouwing Heights, 204–5, 211–12
Wild Oats, 131
Williams, Bill, 41, 56–58, 63, 78, 177, 182, 227
window of opportunity, 100, 161–62, 183, 220
Worst Case Scenario, 143, 147

www.ingramcontent.com/pod-product-compliance
Lightning Source LLC
Chambersburg PA
CBHW061229070526
445#4CB00030B/4049